ELECTRONS and HOLES
put to work in the
SEMICONDUCTOR CHIP

ELECTRONS and HOLES put to work in the SEMICONDUCTOR CHIP

THE 20TH CENTURY INVENTORS of ELECTRONICS and their INVENTIONS

JOHN L PLUMB

Copyright © 2018 John Plumb
All rights reserved
ISBN 13: 9781718185364

PREFACE

We carry in our pockets our smart phones with which we pluck out of the air just about any information we want. It is there available wherever we go at any time of the day, inside or out, in any weather.

Electronics has seen a phenomenal growth at the same time that it has seen a great implosion. Applications have multiplied. Devices have shrunk.

In only a hundred years time, applications have expanded from radio to music reproduction, to television, to computers, to space exploration, to automobiles and to the wireless phone.

Because the electron could not be visually detected, there was little interest in it before the beginning of the 20th century. At the end of the century, not only the electron is too small to be seen, but too, whole electronic parts which control electrons and even circuits which contain these parts are too small to be seen without microscopes. But we are annoyed when our smart phone does not perform as claimed.

Electronics began with discovery of the electron itself. Then it was discovered that the flow of electrons could be controlled. It was found that electrons could be made to flow in one direction and not the other in certain devices. A class of material that is neither a conductor of electricity nor a non-conductor was discovered to exist. And it was found that there are two paths along which this newly discovered material could conduct electricity and that conduction in one path could be independently controlled by conduction in the other. This discovered material is the semiconductor and this discovery made possible an important invention, that of the transistor.

Once the importance of this discovery was realized, many researchers employed in industry, universities and government began to carefully study this phenomenon to understand it. Thus began the enlightenment into the properties of semiconductors, their surfaces and contacts to these surfaces. The effects of methods of growth of the semiconductor, protection of its surface and the nature of the contact to its surface were studied. This enlightenment led to the invention of a very important type of transistor and eventually to the integrated circuit, the microchip, the chip.

The integrated circuit has made possible the innovation of electronic products with which we are now familiar, their low cost which makes them things we can afford, their small size which allows them to be put into a pocket, their speed at which they do their tasks and their reliability upon which we depend.

Inventors in the discovery stage, researchers in the enlightenment stage and entrepreneurs in the innovation stage have protected their work against copying by applying for the patents that were subsequently issued to them. Patents are available from the U.S. Patent and Trademark Office on the Web for reading and downloading by entering the patent number on the number search page at www.patft.uspto.gov. The patent literature is the source of much of the supporting information for the work that follows here.

A glossary is provided at the end of this text. Definitions of terms are shown as they are applied specifically in this document and these terms are italicized when first encountered.

TABLE OF CONTENTS

PREFACE .... **v**

CHAPTER 1 - DISCOVERY **1**
    INTRODUCTION 1
    CONDUCTANCE 3
    THE COHERER 5
    ELECTROMAGNETIC WAVE RECEIVERS 9
    THE WIRELESS TELEGRAPH 12
    DECOHERERS 18
    SELF-RESTORING WAVE DETECTORS 20
    WIRELESS TELEPHON 27

CHAPTER 2 - ENLIGHTENMENT **35**
    THE CRYSTAL RECTIFIER 35
    THE AUDION 38
    CRYSTAL AMPLIFIERS 44
    SEMICONDUCTORS 47
    P-N JUNCTIONS 51
    THE POINT-CONTACT TRANSISTOR 56
    THE JUNCTION TRANSISTOR 59

CHAPTER 3 - INNOVATION 67
THE PLANAR PROCESS 67
THE INTEGRATED CIRCUIT 70
THE MOSFET 76
CMOS 83
THE SILICON GATE 86
FLOATING GATE 88
DRAM 89
MICROPROCESSORS 90
CONTINUING SIZE REDUCTION 92
LATER INVENTIONS 95

CONCLUSION 99
GLOSSARY 101
PATENTS 107
REFERENCES 111
OTHER SOURCES 111
ABOUT YOUR AUTHOR 113

CHAPTER 1 - DISCOVERY

INTRODUCTION

The *electron* is a mysterious bit of physical substance. It is both a particle and a wave. Its particle size is unknown and not well defined but may be less than a millionth of a millionth of a hundredth of a millimeter. That is a decimal point followed by 14 zeros and 1 mm, 0.(00..14 zeros..00)1 millimeter across, or even less than 0.(18 zeros)1 mm, a millionth of a millionth of a millionth of a millimeter. The length of its wave is 0.(7 zeros)1 mm between peaks, or 0.1 nanometer (1 nanometer = 1 millionth of a millimeter). Electrons have an *electric charge* causing them to repel each other and, if free to move, to spread. Electrons are also magnets creating magnetic force, which could be due to their spinning, if they actually can.

Electronics gets its name from the electron. Electronics is a technology which involves electrons, their charge and their flow. Electric charge flow is *electric current,* and in *electronic circuits* it moves in paths. In electronic circuits there are passive components that reduce electric current (as in a *resistor*), resist change in current (as in an *inductor*) or store charge (as in a *capacitor*). The switch and the amplifier are active components. The electronic device we call the *transistor* performs either active function.

At the end of the 19th century, it was discovered that *electromagnetic waves* in the atmosphere could be detected by a crude and slow but simple amplifying switch and that these waves could carry messages. After the transistor was invented mid-century, its size began to shrink dramatically. As more imperfections in material were prevented, the size of individual silicon chips could be increased. As a result of these two trends, the shrinking of transistors and the expansion of chips, the number of transistors on a chip about doubled every two years from 1975 into the 1990s when the 0.35-micron chip generation was introduced. This meant that by this time the finest feature that could be etched into each transistor was 0.00035 mm, since 1mm = 1000 microns.

At the end of the century, Intel introduced their Pentium III Coppermine central-processing-unit *integrated circuit* using a 0.18-micron generation chip (but with the finest feature size a bit longer than 0.18 micron) that contained 28 million transistors. Yes, 28 million. Then, however, in 2001 Intel introduced an upgraded Pentium III with feature size reduced by 30% for their 0.13-micron generation integrated circuit so that the new generation chip contained twice the number of transistors, the result of feature size reduced to 70% of the original, so that feature area was reduced to 0.7 x 0.7 = 0.49, or about half.

In the 20th century electronics became microelectronics and then nanoelectronics. How far this miniaturization can go we do not know. We are far from approaching the size of individual electrons. Some say that the limit will be the distance between atoms in crystalline silicon. In the silicon crystal the atoms are spaced precisely 0.53 nanometer apart. The silicon atom diameter is found experimentally and theoretically to be 0.22 nanometer. An atom of any element has its unique numbers of electrons.

Discovery has been often verification of existing theory. Benjamin Franklin experimented with electric charge in the 1750s and assigned charge as positive or negative, resulting in the electron charge being assigned negative. The electron was not identified as a particle until just before the beginning of the century by J.J. Thomson and his team of British scientists in 1897. It had been named by Irish physicist Stoney in 1891, but the concept of the electron had been proposed by British philosopher Laming back in 1838.

Electrons are moved by electric or magnetic forces. Electromagnetic waves, which include light waves, are electric and magnetic forces perpendicular to each other moving through space at the speed of light. The

Scottish mathematician and physicist James Clerk Maxwell in 1865 theorized mathematical equations governing their motion. The German physicist Heinrich Hertz in 1887 was the first to detect and prove that these waves move through space and that they obeyed Maxwell's equations. It is the unit of the frequency of oscillation of these electric and magnetic forces in cycles per second, *hertz*, or Hz, that is named for him.

Often invention is the result of work by more investigators than the inventor granted the first patent. The Scottish inventor Alexander Graham Bell won a legal battle over the patent of the telephone. In the 1870s, two inventors, he and Elisha Gray rushed their respective designs for a telephone to the patent office within hours of each other. While working with his assistant Thomas Watson in 1875, Bell had entered into his notebook that Watson in the next room had heard him famously say using the device on which they were working, "Mr. Watson - come here - I want to see you". Bell had been educated in the area of speech elocution and correction. Watson provided knowledge of electricity.

Congress had awarded in 1843 to American artist, inventor and publicist, Samuel Morse, funds to construct an experimental telegraph line from Washington forty miles to Baltimore. Morse introduced the Morse code, the representation of the letters of the alphabet by short dots and longer dashes in sound or print. He had successfully exploited in his telegraph receiver an earlier invention - the *electromagnet* - introduced in 1825 by British physicist and inventor William Sturgeon. Sturgeon demonstrated it by lifting nearly ten pounds of iron with a seven ounce piece of iron wrapped with wires connected to a single battery.

CONDUCTANCE

Before reading about electronics we must understand at least the concept of conductance. The ease by which electric charge can move in a material is specified as the material's *conductivity*. Charge can be made to flow in a conductor by a relatively small electric force. The electric current per unit area in a conductor is the product of electric force in the conductor and the conductor's conductivity. For electric force, physicists use the term *electric field*, which is electric force per unit electron charge. We can use the term electric force here.

Current per unit area = electric force X conductivity

The total electric current along a conductor wire is the product of the electric voltage applied between the ends of the wire and the wire's conductance:

Current = electric voltage X conductance

For a metal wire of uniform conductivity, conductance is the conductivity of the metal multiplied by the wire's cross-sectional area and divided by its length.

Conductance = conductivity X area / length

The ease by which charge can be moved by voltage through a length of wire is the conductance of that length of wire. It is this concept of conductance of which we will make use.

At the beginning of the 20th century the telegraph and the telephone were being used to communicate past hollering distance using conducting wires strung between poles. Ships at sea could use signal flags, but only if the ships were within sight of each other, with the aid of an optical lens. There was a great need for the wireless telephone, or at least the wireless telegraph.

The electric voltage generated on Hertz's receiver *antenna* was of sufficient magnitude to cause a visible spark to jump a small gap, a *spark gap*. The spark gap served as a detector and the spark as the detector's visual indicator. The electric charge that jumped the spark gap was however too small to generate the sound or record on paper the dots and dashes of Morse code. At the turn of the century, for a wireless telegraph to produce sound of sufficient strength to be heard by the human ear, amplification by some means was required.

A wireless telegraph would involve the use of a *telegraph key* (a switch) to produce dots and dashes of *alternating-current* (a-c, electric current alternating in direction) in a transmitting antenna. This antenna current would produce electromagnetic waves which in turn would generate alternating-voltage, a-c voltage (electric voltage alternating in polarity) in the receiving antenna. The dots and dashes of the alternating-voltage in the receiver are in 1900 not strong enough to operate a telegraph sounder or to trigger an electromagnetic *relay* to connect a battery to a telegraph sounder.

Some kind of amplifying switch is needed.

In the following sections we will first consider early discoveries involved in the inventions of such an amplifying switch - detectors made from solid material and then liquid detectors. When these detectors were switched on by incoming electromagnetic waves, they had to be switched off before the next dot or dash. We will look at self-restoring detectors, such as crystals and vacuum tubes which also are rectifiers, i.e., pass current more so in one direction than the other. Rectifying devices made possible voice transmission and reception. Three-terminal devices could greatly amplify as well as detect.

Rectification in solid-state devices was not understood, nor was the invention of solid-state amplifiers possible, until the semiconductor was identified and characterized. The transistor which is a solid-state amplifying switch makes use of two semiconductor rectifiers in a single device.

Various innovations followed in new transistors, whole transistor circuits in single crystals of semiconductor, and integrated circuits which are electronic systems on a tiny chip. We will look at some of the significant advancements made by the end of the 20th century.

THE COHERER

The first amplifying switch was a crude device that makes use of an imperfect electrical contact, i.e., a contact with low conductance between two conductors or simply a normally bad contact. Long before 1900, Swedish scientist Peter Munk, in 1844, had discovered a change in the conductance of a collection of metal filings, the particles obtained from filing metal with a file, when the filings are placed near a spark, the *electric discharge* of his Leyden jar (a glass flask lined inside and out with metal foil and filled with water). He had charged the jar from a battery by causing electrons to flow from the foil outside the jar through a wire to the foil inside the jar.

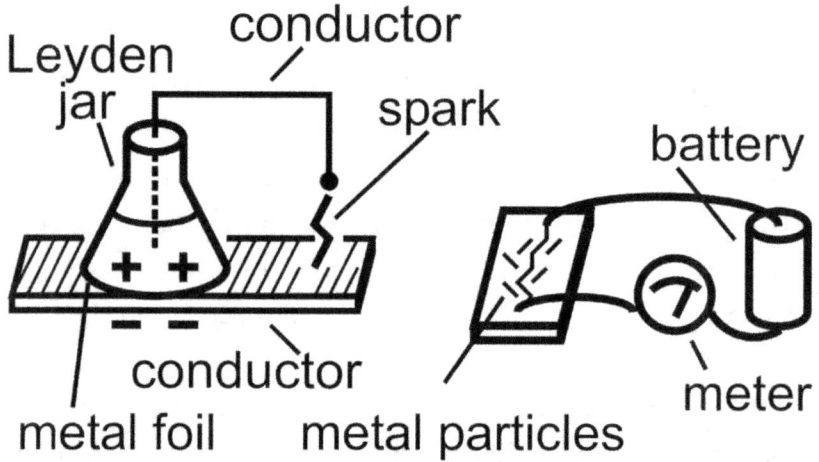

Munk, in an electrical experiment, discovered that the conductance of a collection of metal particles was changed by a nearby spark.

In 1866 the English engineer Samuel Varley used for protection of telegraph lines against lightning strikes, which are electric discharges, a device consisting of two metal spikes inserted into a cavity filled with carbon particles, one spike connected to the telegraph line and the other to the earth. The carbon particles would conduct if the line were hit by a lightning stroke, thus allowing the electric charge in the lightning stroke to flow directly into the earth instead of down the telegraph line.

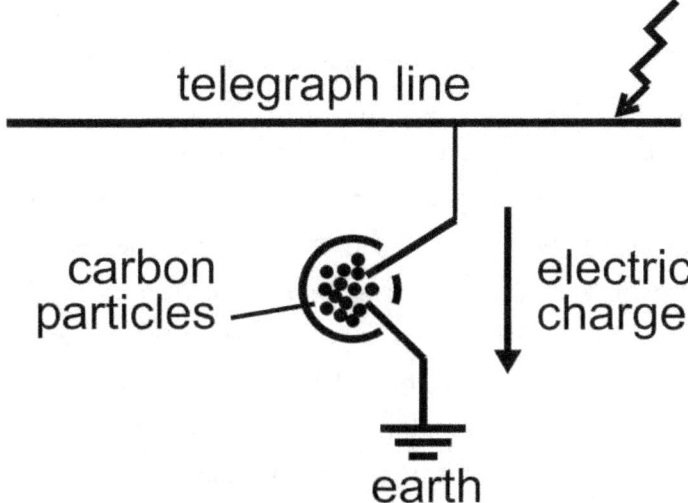

When the telegraph line is hit by lightning, the conductance of the carbon particles increases, thereby conducting the lightning charge to earth.

Metal particles could also be affected by sound. Carbon particles were used in a telephone transmitter invented by David Hughes, a music teacher born in England and raised in the U.S. He returned to London and in 1878 published his work on the effects of sound on the transmitter conductance. He showed that a change in conductance is the result of not a change inside the carbon particles, but instead, the contact between carbon particles. He demonstrated the effect of sound on an imperfect contact by simply laying a metal nail across two other nails connected to a battery and meter.

Hughes also made improvements on a carbon microphone which employed a carbon point or rod resting on a carbon block, an imperfect contact, with which he could detect not only sound but also from far away, electric sparks. He demonstrated this phenomenon in 1880 to members of the British Royal Society, including a Sir William Crookes, who then after seeing the work of Heinrich Hertz, wrote some years later in an article that he had participated in wireless telegraphy. Hughes, not being a physicist and not being aware of any practical use in a wireless telegraph or any other application, did not pursue the work further. He does not mention his findings with electromagnetic waves until 1899, when he does so in The Electrician magazine. There he humbly comments that Hertz's experiments are far more conclusive than his own.

The Italian T Calzecchi-Onesti, a physics professor in Italy, was influential in demonstrating the same effect of remote electric discharge on particles of metals other than carbon from 1884 through 1886. He discovered copper filings between brass plates to cling together and be conductive when he applied a pulse of sufficient electric voltage between the plates. He found that some metal filings could have this same reaction to electric sparks occurring at a distance and that this reaction could serve to detect lightning strikes. His best detector, after many experiments, consisting of a glass tube containing between steel plugs a powder of nickel and silver with traces of mercury, worked well. He was a physicist who understood electricity and magnetism. But electromagnetic waves had only been theorized, and not yet been identified in experiments. Calzecchi-Onesti revealed his findings in 1884-1886 in an Italian journal.

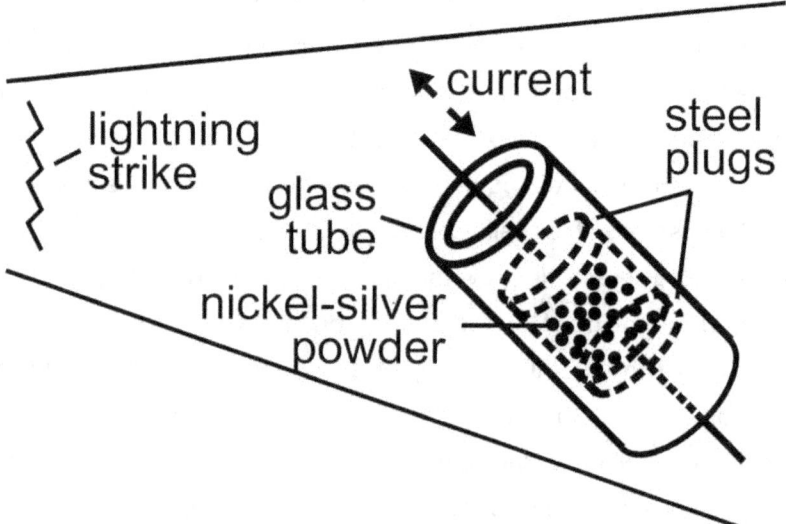

A distant lightning strike was detected by Calzecchi-Onesti in a metal powder.

ELECTROMAGNETIC WAVE RECEIVERS

Lightning strikes generate electromagnetic waves, waves of electric and magnetic forces in the atmosphere. Electromagnetic waves, theorized by Maxwell in 1865 and observed by Hertz, in 1887, were by Hertz purposely transmitted into the atmosphere by connecting a pulse of high voltage to a spark gap through copper wires a meter long. To receive the waves he used a wire loop around which the received waves created a sufficiently high alternating-voltage, voltage alternating in polarity, to cause a spark to jump across a gap in the wire loop.

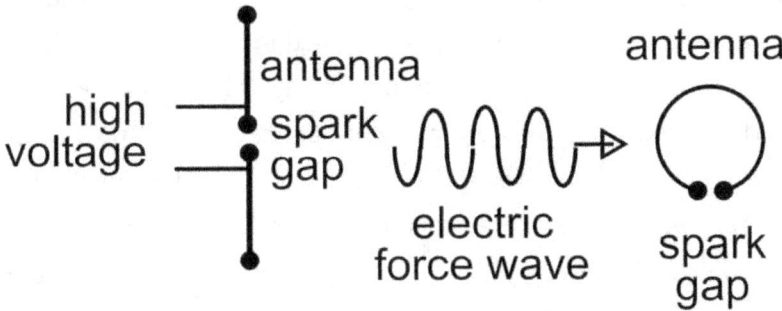

A spark across a gap in Hertz's straight wire antenna generates a wave of electric force in the air which results in a spark across a gap in a loop antenna

Hertz had no idea of the practical importance of this work that would come about in the following century. He wrote that his experiment is "of no use whatsoever....this is just an experiment that proves Maxwell to be right - we just have these mysterious electromagnetic waves that we cannot see with the naked eye. But they are there."

Edouard Branly, inventor, physicist and professor at the Instituit Catholique de Paris, in 1891 published the results of an investigation into the effect of various electrical conditions on metal filings. An electrical discharge as far away as 20 yards caused charge to flow between metal filings between metal plugs in an *insulator* tube, a tube having very low conductance. He found a sharp increase in the conductance of the metal filings. A few sharp blows on the tube brought back normal conductance. This imperfect contact device he described at a British Association meeting in Edinburgh, Scotland, in 1892. That the metal filings in the Branly device may react to electromagnetic waves, as well as electric discharges, was suggested in 1893 by a George Michlin. After Hertz's death in 1894, Oliver Lodge, a physics professor in England, used the Branly device to detect electromagnetic waves.

LODGE

In England, Oliver Lodge at the age of 14 entered his father's business in selling clay to potters and when Oliver reached the age of 22 his father had the financial means to allow Oliver to attend physics lectures and enroll in a local college and then the University of London. In 1877 Oliver was appointed professor of physics at University College at Liverpool, and there he became interested in generating and detecting electromagnetic waves.

On hearing Michlin's 1893 suggestion on the operation of the Branly device as a spark detector, Lodge decided it to be the way to build an electromagnetic wave detector with better performance than Hertz' spark gap. To detect the waves he used an insulator tube containing metal filings and he called it a *coherer*, because in it, the metal filings adhere together and adhere to the metal plates that loosely confine them, therefore cohere. The coherer becoming conductive allows electric charge to flow from a battery and be sensed by a charge meter. He decohered the metal filings, i.e. broke them apart, with a vibrator or a bell in mechanical contact with his coherer.

(How metal filings cohere when sufficient voltage is applied will be better understood in the 20th century.(1) Most switches, relays and connectors grow a thin oxide film on the surfaces of their metal contacts, unless the metal is gold, more so when the contacts are open. When the contacts are closed, electric discharge occurs across the oxide at points where metal particles touch oxide. Electric current flows, temperature rises and metal melts at these points. The points grow in total area until the contacts can carry the current without further melting. Upon cooling, microwelds remain but are easily broken mechanically.)

(In 2008 E. Falcon and B. Castaing in France (2) will experiment with a chain of metallic beads to better understand cohering of metal particles. They line up fifty beads 8 mm in diameter, apply a measured mechanical force between the end beads, and increase an electric current they force through the chain. For low current, the necessary voltage increases. At higher current, the voltage saturates, i.e., does not go higher but the conductance increases and remains high, even when the voltage is reduced. When force is released and the beads rolled to form new contact point, the behavior is repeated. Falcon and Castaing conclude that an oxide film is pierced where metallic beads touch, the size of the contact being about 0.01 millimeter across, and at a voltage per bead of less than a half volt, the temperature rises to over 1000 deg C at these contacts to weld the metal together.)

And then, now at the beginning of the 20th century, an Italian, Guglielmo Marconi, takes advantage of discoveries made by other experimenters who have reported them to the public. One is Serbian-American engineer Nikola Tesla, who has demonstrated wireless telegraphy but has not pursued its possibilities, is instead caught up in an idea of transmitting electric power through the atmosphere. Marconi applies some of Tesla's patents. He becomes an entrepreneur and promoter. The first application of the coherer to wireless telegraphy can be credited to Marconi.

MARCONI

Guglielmo was born into Italian nobility in 1874 to an aristocratic landowner and his Irish/Scot wife. He did not do well at school and was taught English by his mother. He was allowed to tinker in physics and to attend a University of Bologna physics class, where he was taught Maxwell's theories and where he learned of Hertz's experiments. He experimented with electricity and magnetism himself, fabricating much of his own equipment. He built a thunderstorm alarm and one night it is said he woke his mother to show her his alarm, resulting in his father giving him more money for material for experiments.

Now, at the turn of the century in 1900, as an adult, Marconi has set a goal to construct a wireless telegraph. He makes improvements on Hertz's transmitter and wire loop receiving antenna. As did Hertz, to generate the waves he uses a spark gap across which electric charge is made to jump. Marconi's significant invention will be in connecting his spark gap into a circuit loop with a telegraph key and antenna wire to transmit electromagnetic waves for the duration of a dot or a dash. This is space, or wireless, telegraphy, the sending of messages in Samuel Morse's code, through space over long distances.

In 1895 Tesla had been ready to transmit electromagnetic waves 50 miles but his New York City laboratory was destroyed when the building in which was his laboratory burnt down. Tesla gives a few demonstrations that earns him fame and applies for some patents that could have too, but he does not explain much of his work nor does he carry it through to practical application.

We here begin our history of electronics. Telegraph and telephone messages are sent over wires. The wireless telegraph receiver can be said to be the first electronic circuit invented. The electron has been identified, but electronics is not yet in our vocabulary even though it begins with the wireless telegraph in the early days of the twentieth century.

THE WIRELESS TELEGRAPH

Marconi, has traveled with his British mother to England in 1896. There, the technical head of the post office sees the potential in Marconi's wireless telegraph. Marconi applies for patents and for his first U.S. patent he makes application in December of 1896. In this patent Marconi refers to the coherer as a circuit closer. He divulges many details of his system. He has done his earlier work in Italy and would have read about Calzecchi-Onesti's coherer in the Italian journal. In his circuit closer he uses the Calzecchi-Onesti powder, grains of a mixture of nickel and 10% silver with a small amount of mercury. Marconi's powder loosely and partially fills a space between silver plugs inside a small glass tube from which some air has been removed.

Platinum wire is attached to the silver plugs on the inside of the tube and the wire is fed outside the tube to two metal antenna conductors, the length of which are optimized for the length of the electromagnetic waves to be received. Marconi places the antenna conductors, with circuit closer between them, near a reflector which focuses the waves onto the antenna. The other ends of each antenna are connected in an electric circuit loop having *choking coils (inductors* which resist current change), chokes, to block the alternating current, a-c, but allow electric *direct current, d-c*, to flow directly in one direction from a battery through the coil of a *relay*, the battery voltage not greater than 1 1/2 volts.

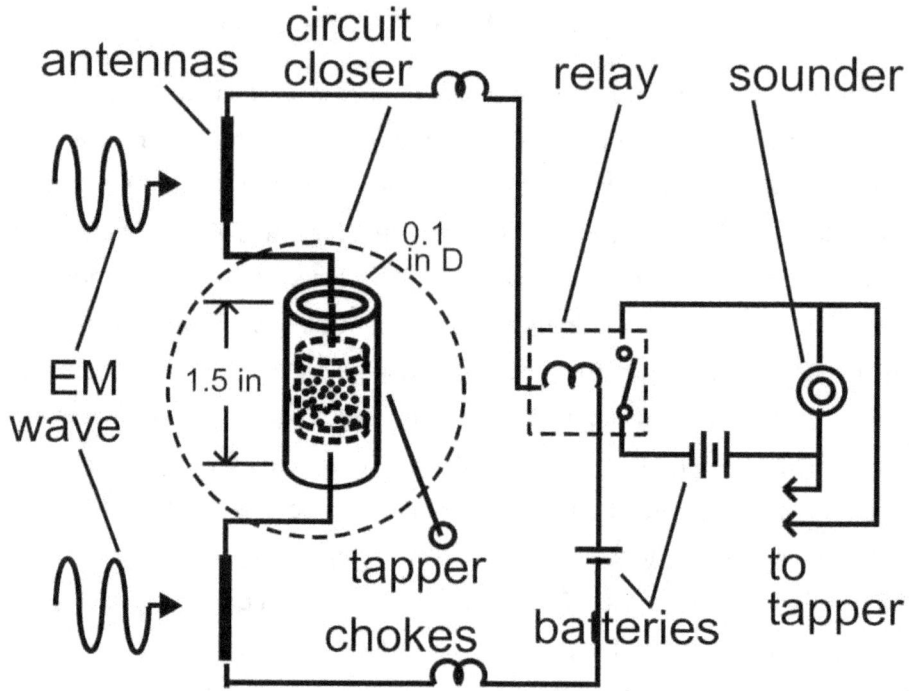

Marconi's wireless telegraph receiver. The circuit closer is a coherer with nickel/silver/mercury powder, silver plugs and platinum wire in a glass tube partially evacuated. Closing of the circuit triggers the relay which energizes a telegraph receiver to sound the dots and dashes and the tapper to open the circuit after each dot or dash.

When sufficient signal is picked up between the two antennas, the circuit closer conductance rises and a sufficient direct current flows from the battery through the circuit closer and inductive chokes to trip the relay. The relay closes another loop carrying higher current from another battery to an ordinary telegraph receiver and also to a device to tap the circuit closer to restore its low conductance. The relay itself effectively amplifies moderate current by switching higher current. The circuit closer (coherer), in effect, amplifies a very small antenna current by tripping the relay.

Thus Marconi has applied an Italian discovery to construct an electromagnetic wave receiver in Italy and then moves to England where he markets it. There with the help of his mother's family fortune and connections, Marconi starts a company he first calls the Wireless Signal

Telegraph Company and eventually the Marconi Wireless Telegraph Company. He thus is the entrepreneur to bring wireless telegraphy to the world. Marconi's patent has been issued near the end of the 19th century in July of 1897. In his patent application he attempted to capture the market by making all of 47 invention claims which cover the coherer containing the mixture of metallic powder, the antenna plates with the choking coils, the relay, and the tapper.

On February 1, 1898, Oliver Lodge, the professor who has lectured on electromagnetic waves and detection of them using his coherer, has decided to jump into the competition with Marconi and files for a patent that describes similar but different electrical transmitting and receiving circuits. In the receiver his coherer is similar to that of Marconi, but platinum wires are brought close together directly into filings of iron. A tapper is not always necessary to restore the coherer's normally low conductance.

In one form his receiver has a coil (*inductor*) between two *capacity* (*capacitance*) areas (antennae) for *syntony* (*resonance*) between the coil and antennae at the oscillation frequency at which a similar circuit in the transmitter is *resonating*. The coil is also in a circuit loop with the coherer, and the coherer also in a loop with a battery, without a relay, and, to sound the dots and dashes, a telephone receiver. In another form, the coil connecting the antennae is surrounded by another coil so that the two coils are *magnetically coupled*, resulting in what he calls an *induction coil* (*transformer*) which magnetically steps up the voltage in the second coil, this second coil in the circuit loop with the coherer. The patent is issued in August of 1898, and in it Lodge claims invention of the combination of induction coil and coherer loop without needing a relay. It is a receiver that requires fewer parts than does Marconi's receiver.

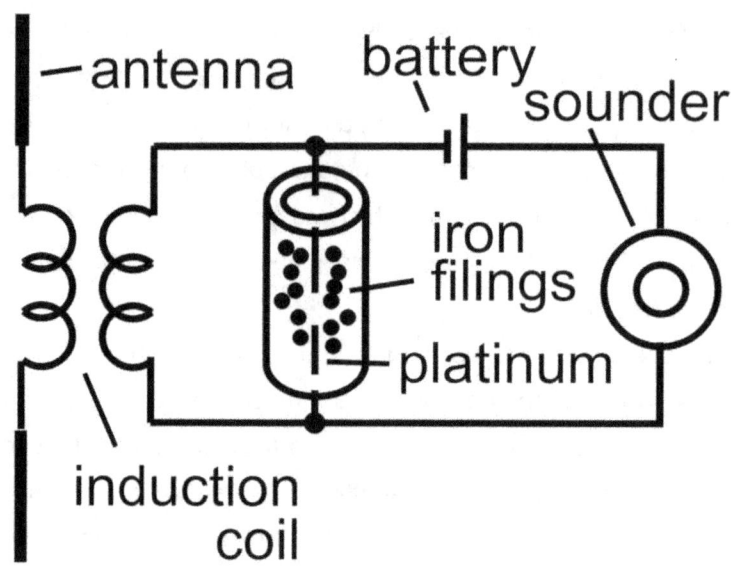

Lodge's radio wave receiver incorporating an induction coil to resonate with the antenna at a transmitter frequency and a coherer with platinum wires extending into iron filings

Then in January of 1899, Marconi files for two more patents on improvements to the coherer in his system. In one patent, issued in June, Marconi takes Lodge's receiver which he has obviously seen described in Lodge's patent and modifies it by connecting the primary coil of an induction coil (transformer), similar to Lodge's induction coil, but between an antenna and the earth. He has discovered that reception is greatly improved by connecting the receiver between one antenna and the earth. This is a patentable discovery in itself.

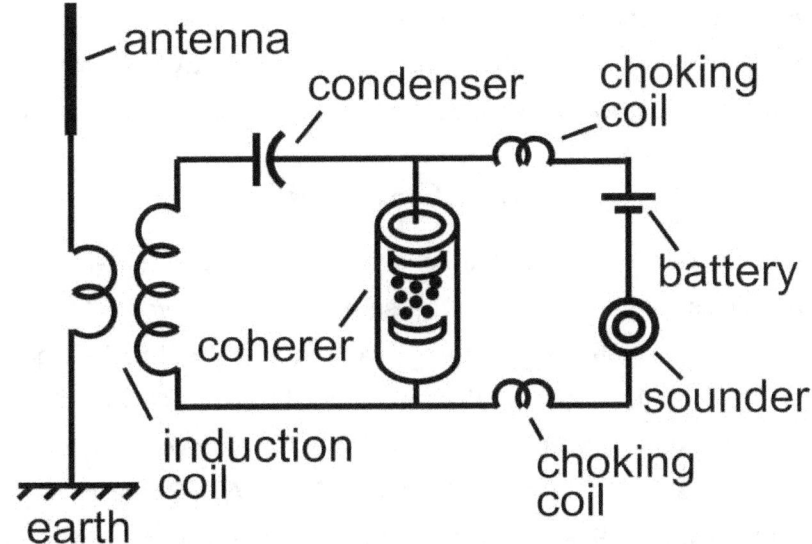

Marconi radio wave receiver with earth connection, condenser (capacitor) to block battery direct current and choking coils (inductors) to block antenna alternating current

Marconi divulges the details of the condenser and induction coil construction including the number of turns of wire and thickness of the wire in the induction coil and the capacitance of the condenser chosen for syntony (resonance) with the wavelength of the waves transmitted.

In 1899 Marconi reports by wireless telegraphy the results of the International Yacht Races off Sandy Hook, New Jersey. This same year he sets up his company in the United States.

At the beginning of the 20th century, in 1901, Lodge forms a syndicate with other wireless telegraphers to contest Marconi's patents. Lodge claims that Marconi has used ideas already patented by others. It appears that Marconi has purposely applied the patents of others and made his own as secure as he can.

That year Marconi goes to Newfoundland to receive transmission from England. His company takes the lead in a race against competitors in the U.S. and in Germany to control wireless telegraphy. He takes out patents on every device his company makes. He hires John Ambrose Fleming, an

English electrical engineer, to be his advisor and to work on improving the coherer. Fleming will later make a major contribution. By 1903 the Marconi Wireless Telegraph Company will have a virtual monopoly on the business.

DECOHERERS

The necessary tapping of the coherer after each dot or dash detected obviously limits the speed at which messages can be received. A number of inventors file for patents in 1899 and following years and make significant improvements on the wireless telegraph, many having to do with turning off the coherer, i.e., decohering it, beginning with a novel construction of coherers in a patent awarded in February of 1900 to A.F. Collins of New York for the American Wireless Telegraph and Telephone company, a would-be competitor against Marconi. In Collins' receiver, the metal filings in the coherer lie between beveled contact plugs that form a trapezoidal-shaped space that widens upward. An electromagnet on top of the coherer is energized to pull the filings into a wider space to decohere the coherer.

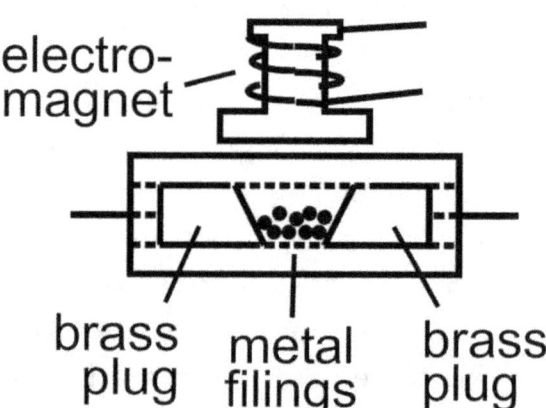

Collins' coherer with beveled plugs and electromagnet to raise filings into wider space to restore low conductance.

Another wireless telegraph company has sprung up. In October of 1901, G.W. Pickard of Boston for Consolidated Wireless Telegraph and Telephone files for a patent for a coherer that is a small modification to Collin's coherer.

It has, instead of beveled contact plugs as described in the Collins patent, but rather plugs with a concave depression at their ends or plugs with a cone-shaped depression. Both shapes form circular cavities which widen toward the center rather than the top to insure both cohering and decohering of filings for any position of the coherer tube, thus requiring less precision in plug orientation.

Steel is found to work well as coherer material that decoheres more easily. In his patent application in May of 1899, Eugene Ducretet in France has used hard-tempered steel powder between movable rods. His patent is issued in January of 1901. The accuracy in breaking the contact to decohere is found to be greater the harder the metal used for the metal plugs. In August of 1900 Adolph Koepfel in Germany has applied for a U.S. patent on a coherer having hardened steel or cast iron plugs outside steel, silver or other metal powder. (Hard metal does not deform as much so does not create as large a microweld where particles cohere.)

In September 1902, Branly, having been recognized as being the principle inventor of the coherer, files for a patent on a wave receiver using for a coherer the imperfect contacts between significantly oxidized blunt steel points and polished steel plates, which he now calls a radio-conductor since he now recognizes it to be a conductor the conductance of which is triggered by radiated electromagnetic waves. The oxide blocks current flow when waves are not being received. Oxide likely plays a role in the Ducretet and Koepfel inventions. Branly reported that only a very slight mechanical shock was required to decohere. Branly's patent, however, will not be issued until August of 1905.

(Branly's calls his coherer a radioconductor, but radio, voice transmission, will not use any type of coherer, but rather a *rectifier*.)

Also in 1902, Lodge "to dispense with the use of imperfect contacts.....which are subject to variation of sensitiveness and require frequent adjustment or renewal" files for a patent on a wave detector having a metal point to push a thin oil film into a pool of mercury. This detector makes use of electric breakdown, arcing, across thin insulator film as will be believed to occur across oxide film on metal filings. At a voltage of little more than one volt, the oil film begins to conduct. This wave detector is not a coherer, but instead makes use of conductance that depends on the electric force applied across a thin liquid insulator instead of between metal particles.

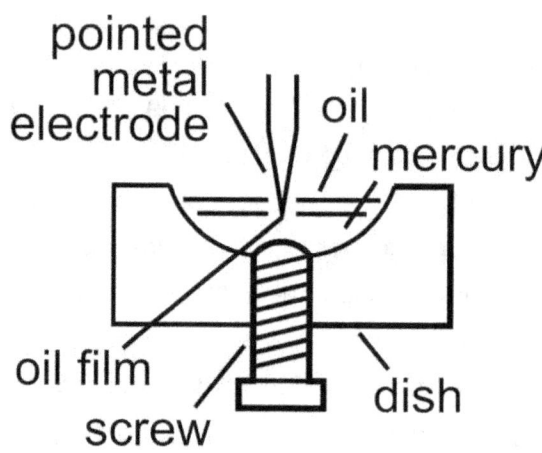

Lodge wave detector employing a thin insulating oil film with conductance increased by a received electromagnetic wave.

But the oil film must lose conductance at the end of every dot and dash of Morse code. In two embodiments of his device, Lodge causes the pointed metal electrode to be raised in one and in the other continuously vibrates the device to restore non-conductance.

SELF-RESTORING WAVE DETECTORS

A coherer that does not require tapping is needed. Carbon parts and particles have been found to act as coherers that self-decohere, i.e., become conductive when high voltage is impressed and spontaneously lose conductance when high voltage is removed. This solution may avoid the microwelds. In September 1900, Isodor Kitsee of Philadelphia has been awarded a patent for a receiver employing an imperfect contact between two carbon buttons. Incoming current pulses from the antenna increase the conductance of the carbon contact sufficiently to actuate a relay, but not enough to create microwelds, so when the pulsations cease the imperfect contact restores itself.

Carbon particles loosely placed around and between steel balls are used as a detector in a receiver described in a patent application filed February of 1901 by an H Shoemaker. He claims that no decoherer is necessary in his receiver in which the detector is connected directly between his antenna and the earth but connected also in a loop in a unique circuit with batteries in which a Morse code dash is received as two dots in rapid succession. Shoemaker is awarded over 20 patents from 1901 through 1903 for inventions, all of supposedly improved wireless telegraphs, but all using coherers in the receivers.

A sharp carbon pencil slowly rotated against a flexible metal spring is used for a self-restoring wave detector by Joseph Murgas in 1903 in Wilkesbarre, Pennsylvania. His telegraph system also departs from transmitting and receiving dots and dashes of two time durations and instead adopts transmitting waves oscillating at two frequencies. He claims that such signals may be transmitted at higher rate and the two signal frequencies may be more easily distinguished from each other by a telephone in place of a telegraph receiver. But by the 1903 patent issue date, other types of wave detectors, mainly ones using a liquid or semi-liquid, have been invented.

Lee de Forest, an engineer in the research department of Western Electric, manufacturer of telephones, devises a wave detector, using an idea from a German publication, and calls it a responder. Essentially, his responder consists of two plates of metal separated by a thin semi-liquid substance in which metal bridges form. His major contribution however to wireless telegraphy will later be a vacuum tube.

DE FOREST

Lee de Forest was born in 1873 in Council Bluffs, Iowa, to a Congregational Church minister, Henry de Forest. When Lee was six years old, Henry was called to be president of Talladega College in Alabama, a college founded for freed slaves. It was an unfriendly situation for Lee, not liked by either the black or white children. There Henry established a grade school, in which Lee was educated and in which he excelled, but the most important part of his education was his own reading. He had access to the college library and carpentry shop, and his father subscribed him to the magazine Youth's Companion. He began to build and invent, and he kept a diary from which we know much about his early life.

Henry had wanted Lee to study for the ministry, but he reluctantly let Lee attend Sheffield School, a technical school with connections to Yale University, in Connecticut. Lee was 20 years old. There he invented and built things like a steam condenser, trolley system, pants creaser and ear cleaner. But no manufacturer took notice of his ideas. He designed a draftsman's compass, typewriter, and puzzles. He chose to study mechanical engineering, but his interests would soon turn to wireless telegraphy.

After three years at Sheffield his father died, and Lee's mother moved to Connecticut and opened a rooming house. That year Lee got his bachelor's degree from Yale University and began electrical studies. He began reading Hertz's work in German and wrote a research thesis that extended Hertz's experiments on length and frequency of electromagnetic waves. In 1899 Yale conferred upon him a Ph.D. degree. He had studied the writings and patents of the great electrical engineers of the day, namely Edison, Westinghouse, Tesla and Marconi.

De Forest accepted a position at Western Electric in Chicago and after being assigned to work with telephones, he spent time on the side developing a wave detector of electromagnetic waves, his oscillation responder, for wireless telegraphy. After seeing a Marconi demonstration he begs for a job with Marconi but gets no reply. He takes a job with American

Wireless Telephone Company in Wisconsin, where after deciding the company's wireless telegraph wave detector impractical, he substitutes his responder.

For a responder he experiments with many liquid conductors, settles on a paste of glycerin and a lead oxide, and calls the paste simply "goo". When battery voltage is applied between electrodes, metal particles leave the negative electrode, he claims, and form paths for electric current. (Likely positive metal ions are attracted to the negative electrode and so form metal paths for electric current.) These paths are disrupted by alternating-current from the antenna, and the paths are immediately reformed by battery current when the antenna current ceases The responder conductance decreases when oscillations are detected rather than increase, and so the responder is called an anticoherer type. The oscillations can be heard with a telephone receiver or the decrease in battery current can be made to actuate a relay, which in turn actuates a telegraph receiver.

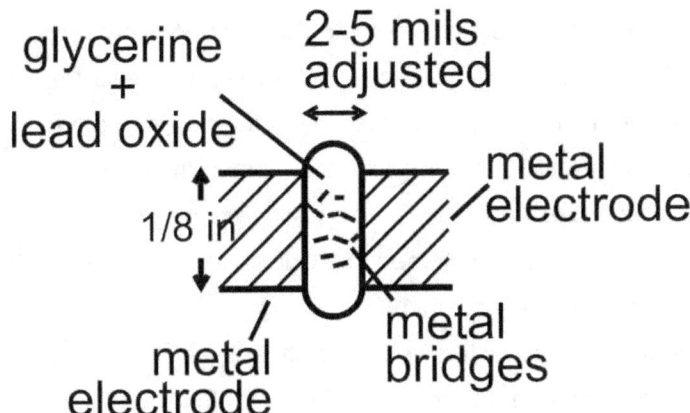

De Forest's responder in which metal bridges form when battery current flows and break when also antenna oscillating current flows

De Forest refuses to divulge the construction of his wave responder. He leaves the company and moves from Wisconsin back to Chicago. With former Western Electric engineer Ed Smythe, who financially supports him, and Armour Institute professor Clarence Freeman, who helps with experiments, he works to perfect the responder and applies for a patent in July of 1901. In December 1902 he and Smythe are awarded the patent. Like Marconi they try to cover all bases. Their 65 claims include electrodes with roughened faces and the ingredients of the paste in which bridges form and break.

The three form the De Forest Wireless Telegraph Company. In six months the company has installed seven stations around the country. For the U.S. Navy they install five stations. De Forest tries out this receiver with his responder on Lake Michigan and attains a range of 4-5 miles. Coherers are widely used but are soon replaced by de Forest's "goo" responder or Reginald Fessenden's hot-wire barretter.

FESSENDEN

Reginald was born in Quebec, his father a minister who moved his family to various congregations in Ontario. He was an accomplished student and studied at Bishop's University while teaching younger students at Bishop's College School on the same campus. At 18 years of age he left Bishop's, without completing the requirements for a degree, and landed a job with Thomas Edison in 1886 as an assistant tester in laying electrical cable underground in New York City. He was soon assigned to Edison's new facility in West Orange, New Jersey, but Edison ran into financial trouble and laid off most of his employees in 1890.

Fessenden then accepted an appointment as a professor in a new electrical engineering department at Purdue University. He helped the Westinghouse Corporation install lighting at the Chicago 1893 Columbian Exposition, and then George Westinghouse recruited him to become the chairman of a new electrical engineering department at Western University in Pennsylvania, (which would become the University of Pittsburgh). There he began experimenting with wireless communication and by 1899 Fessenden had installed a wireless system between Pittsburgh and Allegheny City, PA.

At the turn of the century, Fessenden has been hired by the US Weather Bureau to set up radio stations to transmit weather information. There from December 1999 to May 1901 Fessenden files for patents on various wave detectors utilizing deflection and attraction between freely-moving current-carrying coils.

Now, in a May 1901 patent application, Fessenden is using for a detector a coherer and is tuning receiver input loops to the transmitter frequency. He is also using an *electrolytic cell* to just balance the battery voltage so that no battery current flows when there are no antenna oscillations. He will later recognize that the electrolytic cell plays another important role.

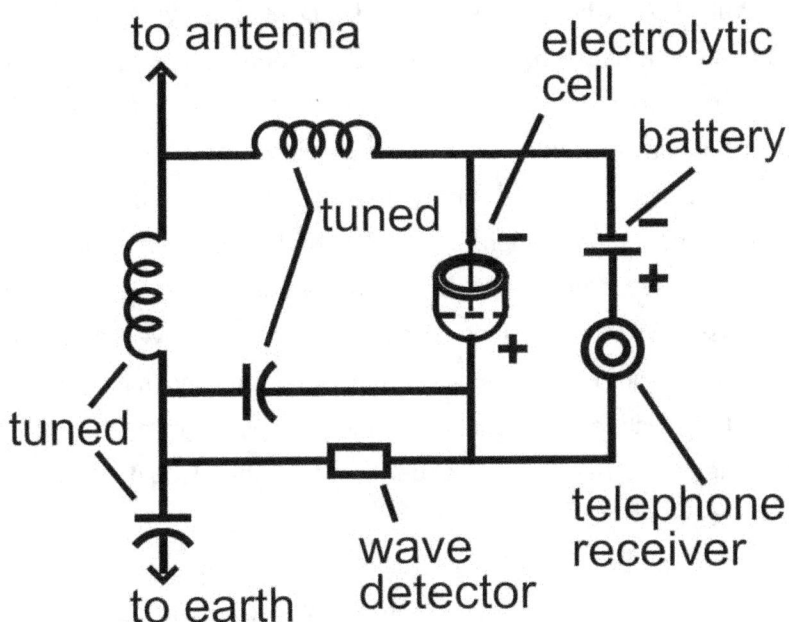

Fessenden's radio receiver employing tuning components and an electrolytic cell

In another May 1901 patent application Fessenden has proposed that if at the transmitter the voltage source that supplies energy to the spark gap is an alternator that generates alternating (a-c) voltage, the sparks may occur at the frequency of this a-c voltage. The transmitter can be tuned to its antenna to increase antenna current at that frequency and the receiver at the receiving station can be tuned to select that frequency. Fessenden's patent is issued in August 1902, but he does not have yet the alternator.

Fessenden then has success with a detector he calls a hot-wire *barretter,* a barretter being a conductor of which the conductance varies with temperature. It is much like a light bulb with a very fine filament that is heated by incoming waves. It works as an anticoherer in which, like in the Lodge and de Forest wave detectors, the conductance is reduced by incoming waves. In a patent, for which he files in June of 1902, he describes a silver wire having a platinum core 3 mils (0.003 inch) in diameter from which he dissolves the silver over a length of 1 mil. This filament is sealed in a glass bulb from which air is pumped. Due to its very small size the exposed platinum heats very quickly when antenna current passes through it. Its electrical conductance decreases for each Morse code dot or dash of current oscillations received and, in between, it increases.

Within each dot or dash a number of sparks occur across the transmitter spark gap. Each spark generates oscillations the strength of which rapidly decays. At the receiver these oscillations are heard in a telephone receiver as just noise. Without this noise Fessenden now believes that his liquid barretter could reproduce transmitted voice, if for the transmitter a generator could be built with a high enough rotational speed, rpm, to produce an electric voltage alternating at a frequency higher than would be audible in a telephone receiver.

Fessenden proposes that if the strength, *amplitude*, of the transmitted wave could be made to vary at the frequency of human voice, a wireless telephone would result. Since voice into space will be radiated, the term wireless telephony will be shortened to *radio.*

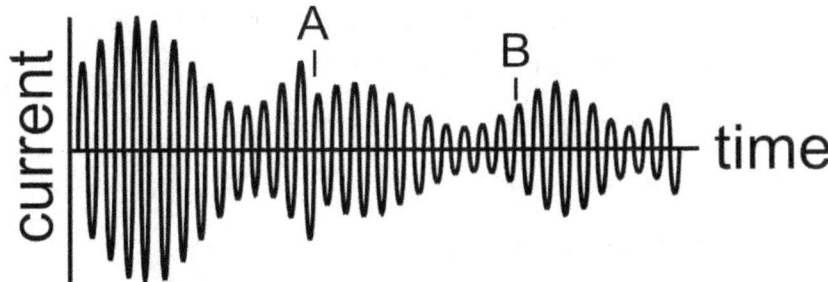

A representation of antenna alternating current modulated by a sound wave. Here the sound strength is reduced at point in time A and then the sound frequency is increased at point in time B.

This is *amplitude modulation, A.M.,* radio (3). Two wealthy Pittsburgh men see in Fessenden's idea a great opportunity and capitalize a National Electric Signaling Company which the three organize.

(In a pure alternating current, the current swings the same amount one direction as the other and the average current is zero. A wave detector will not detect the sound frequency unless the average current is not zero and the detector can follow the average at sound frequencies. If the detector allows more current in one direction than the other, it is said to *rectify*.)

THE WIRELESS TELEPHONE

In 1903, the year that Wilbur Wright flies his and his brother Orville's motorized heavier-than-air aircraft at an altitude of fifteen feet for nearly one minute, Fessenden files for and in the next month is issued a patent describing a wave detector using a liquid and offering a bigger change in conductance upon responding to waves than does his hot wire detector. It employs two fine platinum wires the ends of which are separated by a liquid conductor, preferably nitric acid. According to Fessenden, conductance near the ends of the platinum wires, which is normally low, is raised through heating of the liquid by the high-frequency antenna current.

In another embodiment of his patent Fessenden describes a detector having only one platinum wire point dipped into acid in a cup and the other platinum wire immersed in acid at the bottom of the cup. The cup and the wire point are connected into a receiver circuit.

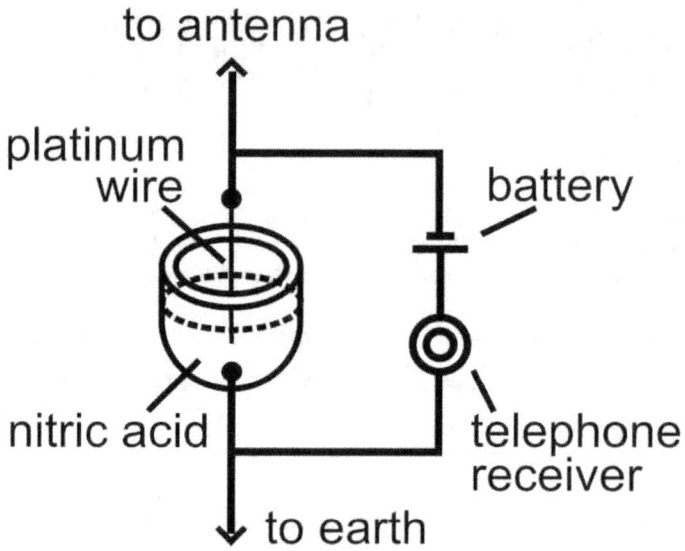

Fessenden's receiver employing his liquid detector having one platinum wire tip immersed in nitric acid.

In his patent Fessenden states that the detector is not rectifying, but that at higher battery voltage he observes bubbles forming at the wire tip, indicating some electrolysis, which is possibly enhanced by the heating of the acid, and hydrogen gas coming off from the nitric acid. He also states that the detector performance was better with the platinum tip made negative by the battery, possibly due to less gas present around the wire tip. (More likely, electrolysis is enhanced at this polarity and some rectifying is occurring.)

Fessenden now acknowledges that in his earlier patent on a receiver employing a coherer and an electrolytic cell, the use of the electrolytic cell to oppose the battery in the absence of electromagnetic waves was a "method .....for rectifying the alternating currents produced by the radio electromagnetic waves," but in this receiver above "this effect does not occur." He also states that the hot-wire barretter rectifies. (This does not seem likely. It is likely he, in 1902, does not have the means to clearly identify rectification.)

In June 1903 de Forest, after Fessenden's patent has been issued, applies for a patent in which he varies the frequency at which sparks are created by using an a-c generator to vary the gap length of his spark gap. He varies the frequency in slow cycles in a transmitter to create a warning sound like a fog horn in a remote receiver. He claims that any wave detector can be used, but did not name any that he used successfully.

On a visit to Fessenden's laboratory, de Forest finds Fessenden using for wave detection the liquid detector. De Forest and a former employee of Fessenden design their own such detector, using two platinum wires with their extreme ends in acid, but sealed in a glass tube, and they call their wave detector a *spade detector*. They claim it to be different from Fessenden's liquid detector and sell it at a price below that of Fessenden. It will become the basic detector for the de Forest system. But later de Forest will be found to have infringed on Fessenden's patent.

Fessenden, using sparks generated and tuned at an alternator frequency of 20,000 cycles per second, hertz or Hz, at the transmitting station and the antenna circuit tuned to that frequency, demonstrates radio before a group of engineers who agree that the voice articulation was satisfactory in signal transmitted 25 miles. He claims that the wave detector is not rectifying the signal, i.e., not conducting more current in one direction than the other. He advertises wireless telephone sets for sale, but later admits the system is not perfect.

So that year, 1904, Fessenden orders an alternator capable of generating an alternating current, a-c, at a frequency of 100,000 hertz, from General Electric in Schenectady, NY. The design work is assigned to Ernst Alexanderson, a recently hired engineer, who comes up with a design for an alternator with a stationary iron armature between two rotating disks. Fessenden requests that the alternator be built without iron, the iron causing spark oscillations to rapidly decay. Alexanderson submits a design to

Fessenden for an alternator with a wood armature but reiterates his opinion that iron is preferable. In 1905 he reports to Fessenden that a machine using wood instead of iron that Fessenden had constructed elsewhere is defective and not worth more time. But finally in October of 1906 Alexanderson will report he has achieved 50,000 Hertz in an alternator without iron. Fessenden's insistence will be rewarded.

In December 1905 Fessenden's competitor, de Forest, applies for a patent describing various means to vary current from a generator delivering alternating current at 750 hertz to a transmitting antenna using sound from a sound amplifier such as a megaphone. A means to vary the current can be a carbon microphone. But, as in his patent application the previous year, the patent does not cover a receiver to reproduce the sound.

Fessenden's company has abandoned a radio station on Chesapeake Bay and has built one at Brant Rock on the coast of Massachusetts. The antenna is a 420-foot high 3-foot diameter metal tube assembled in 8-foot sections. He has a similar station built in Scotland, and in January 1906 the two stations exchange telegraph messages.

By the Fall of 1906, Fessenden has his alternator delivering alternating voltage at up to 100 KHz. In November voice messages were sent between the Brant Rock station and a test station at Plymouth, Massachusetts, and the voice message transmitted from Brant Rock is heard at the Scotland station. The tower at the Scotland station is blown down in December and is never rebuilt.

On December 24, 1906, US Navy ships at various locations over the Atlantic Ocean and the West Indies with radio receivers equipped with Fessenden's liquid detectors receive from Fessenden's Brant Rock station the first A.M. radio broadcast, consisting of a program of Christmas music.

In the meantime, de Forest's factory in Jersey City works overtime to fill orders for wireless transmitters and receivers and he opens a laboratory in New York City. However, in 1906 the federal courts declare that de Forest's receiver detector, his spade detector, is an infringement on Fessenden's patent for the liquid detector. Fessenden will be given the credit for the introduction of radio.

The year 1907 is not a good year for de Forest. In July de Forest makes ship-to-shore transmission by radio telephone for a yacht race on Lake Erie. He interests the Navy in his radiotelephone, which places an order for 26 sets, but they are declared unreliable. His company sets up stations along the Atlantic coast and the Great Lakes but they prove unprofitable. De Forest's backers force him out of his own company. The company holds onto his patents, but de Forest has quietly begun to develop a completely new idea that he has not yet divulged.

He has a wire grid placed between the filament and a plate in a light bulb, with a wire lead from both the grid and the plate running out of the bulb. He will have truly his own invention which will later bring him fame, the wave detector and amplifier which is the vacuum triode.

THE VACUUM TUBE

Isodor Kitsee has been awarded in 1895 a patent for the first vacuum tube used in wireless telegraphy. It actually uses the glow between two plates in vacuum inside a glass tube as a visual indicator of the dots and dashes of a message received in Morse code

Because Edison's light bulb, with its hot filament and a plate electrode added, conducts current only in one direction, it is a rectifier. It will become an important wave detector that is useful for receiving voice messages as well as telegraph messages. When high-frequency a-c voltage appears across this detector, the current between filament and plate is rectified and direct current flows for swings to one polarity of the high-frequency voltage. If the frequency is sufficiently high that telephone receivers cannot follow them, then the direct current in the telephone receiver at the receiving station will follow the average of the rectified current swings and the voice signal will be understood. The first to apply this device to radio wave reception is John Fleming, the consultant for the Marconi company.

FLEMING

John Fleming was born in England in 1849, eldest of seven children and son, like Fessenden and de Forest, of a church minister, he in the Congregational Church. John was a devout Christian who once gave a sermon on evidence of the resurrection. Finances forced him to alternate study with teaching. He was appointed the first professor of electrical engineering at University College of London and consulted for Thomas Edison, for the Marconi Company and for others.

It is 1904. Fleming has studied and published his findings on the carbon deposits seen by Edison in his light bulb. For a wave detector for Marconi, he modifies Edison's light bulb having an added small plate. He makes the metal plate a cylinder surrounding the filament. In his patent he calls his detector an oscillation valve. (In England vacuum tubes will be called valves.)

This oscillation valve is connected in a loop with one winding on an induction coil (transformer) and a galvanometer (charge meter). The other winding is connected between an antenna and earth.

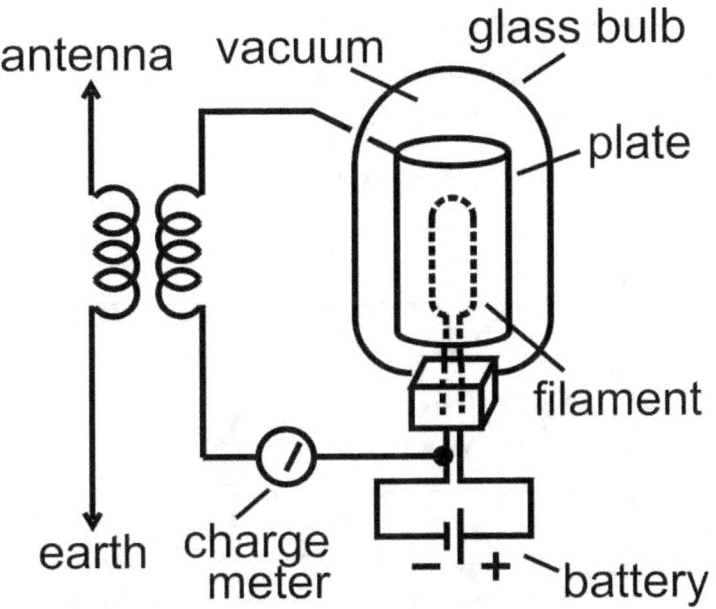

Fleming's oscillation valve (vacuum tube) in a circuit in which current flows in one direction between filament and plate and through a galvanometer (charge meter) when electromagnetic waves are received at an antenna. Current from the battery heats the filament to emission temperature.

This is a perfect detector for A.M. radio. When electromagnetic waves are picked up by the antenna, high-frequency alternating current flows in one coil and rectified current, in one direction only, through the other coil, oscillation valve and charge meter. This current moves the galvanometer needle indicating the presence of waves in the air at the antenna. Fleming files for a British patent, and then in April 1905 for an American patent. In October 1905, de Forest sends an assistant to Henry McCandless in New York City to duplicate Fleming's valve having a nickel plate connected to a wire which protrudes from the top.

Fleming's patent is issued in November of 1905. Soon after, in January 1906 and again in May, de Forest files for patents for radio receiver circuits using a detector that he claims to be an improvement on Fleming's oscillation valve and he calls it a gaseous oscillation responsive device. Like the Fleming valve his detector consists of a filament and a plate in a bulb. Like

Fessenden's receiver that used a liquid detector, his oscillation responsive device is connected between antenna and earth and also in a circuit loop with a telephone receiver and a battery.

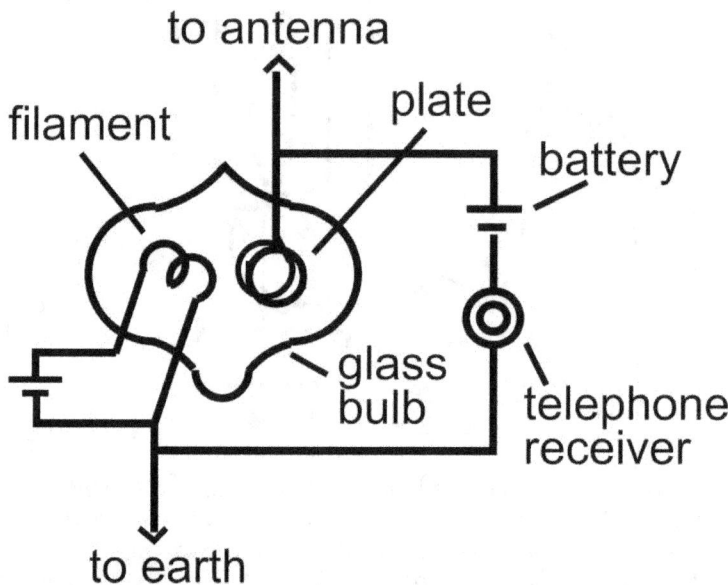

De Forest's radio receiver employing his oscillation responsive device consisting of a heated filament and conducting plate inside a partially evacuated glass bulb

Rectifying devices will become essential in radio receivers and in many other electronic applications. As understanding grows of the nature of conduction in solids and at contacts between solid materials, rectification will be accomplished in smaller and smaller two-terminal devices and will be become a necessary property in three-terminal devices designed to control current, to amplify it or to switch it.

CHAPTER 2 - ENLIGHTENMENT

THE CRYSTAL RECTIFIER

R. Braun, before 1883, in working with crystalline minerals found in nature, applied point contacts to some mineral crystals and noted that the conductance can depend on, not only the condition of the surface of the material, but also on the direction, *polarity*, with which voltage is applied between the material and a point contact.

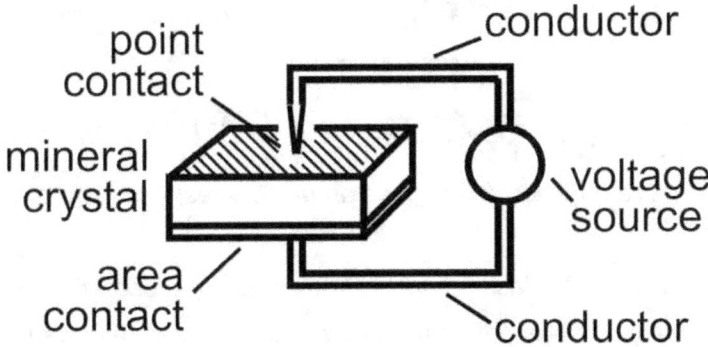

An electrical point contact to the surface of a crystal of mineral

Jagadis C. Bose, Calcutta, India, at the turn of the century in 1901 has filed for a U.S. patent on his discovery that various materials conduct an electric current that does not increase proportionately with an applied voltage. In some the conductance increased with voltage, as it does between carbon contacts, and in some it decreased, as it does in Fessenden's barretters.

One material that Bose found to have a conductance that increases with voltage is galena, a lead ore that is a lead-sulfur compound mined, for one place, in Galena, Illinois, on the Mississippi. Bose is apparently unaware that a small contact to lead sulfide can result in conductance greater in one direction than the other. It rectifies and this property becomes an extremely useful property. Galena and compounds of sulfur with iron as radio wave detectors become popular in crystal receivers in 1906. Galena proves to be the most effective one. The crystal detector is not expensive, amateurs use it and it becomes standard for home-made radios.

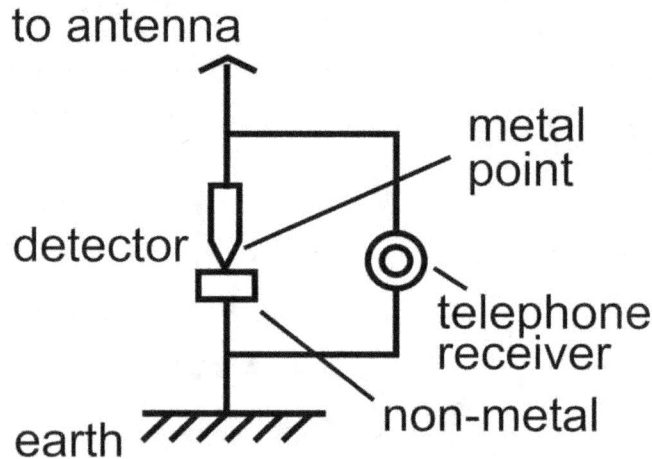

Simple wave receiver which gives an audible response in a telephone receiver.

William Hogg of Washington D.C. files for in February 1903 and is awarded in June 1904 a patent for a wave detector that employs the element selenium. His invention he writes "relates to improvements in receivers in which with imperfect contact there is an automatic decohering action." He rubs selenium onto the surface of a metal wire, heats the wire to form a crystalline mixture and electrically contacts this crystalline material with a steel needle.

Another sensitive detector material is carborundum. In 1906 retired Army general Henry Dunwoody, now with the De Forest Wireless Company, files and is issued a patent for a crystal radio receiver using for "an improved wave responsive or detecting device" a compound of carbon and silicon, carborundum, a high-temperature-resistant hard poorly-conducting non-metal. Dunwoody claims that the carborundum could be contacted by any of various means - winding a wire around, positioning between plates, touching another piece of same or dipping in an electrolyte. Like with a galena detector he claims that, with sufficient antenna, a battery in a telephone receiver loop is unnecessary.

The common belief is that in such contact detectors the radio frequency signals heat a small area which generates a current. G.W. Pickard, who patented coherers with plugs having cone and concave-shaped ends, working at his home in Amesbury, Massachusetts, experiments with iron oxide, lead oxide and selenium as the non-metallic element contacted with the point of a good conductor. In an article in the London Science Abstract he reads of a new product made in an electric furnace - fused silicon. Silicon is hard to find, but when he has received a sample from a Henry Potter of Westinghouse, he finds it to outperform any of his previous materials.

PICKARD

Greenleaf Whittier Pickard, born in Maine in 1877 and named after the poet, was educated at the Lawrence Scientific School, at Harvard University and at the Massachusetts Institute of Technology. In the summer of 1899 he received a grant from the Smithsonian Institution to experiment with wireless antennae raised by kites at the Blue Hill Observatory in Massachusetts. In 1901 he went to work for American Wireless Telegraph and Telephone in Boston. That year he set up to report the America's Cup yacht races, in competition with the Marconi and De Forest companies.

In 1902, for his company, Pickard began investigating wireless telephony, radio communication. Using a microphone diaphragm to change the gap length of a transmitter spark gap to send voice signal and a carbon-steel detector to receive, he claimed to have sent and received speech, but only a short distance. He was assigned other work, but continued to investigate radio at home. He tried steel needles in contact with oxidized steel, first an old scissors, then with iron oxide on steel, with improved reception. The best materials all had low conductivity. He set up a home receiving station in Massachusetts where for three years he tested many materials having low conductivity.

Pickard then learns of silicon, but can find it only as a powder, until August 1906 when he obtains a sample that outperforms all the other materials he has tested. He, his patent attorney Philip Farnsworth and his salesman Col. John Firth, who has close ties with the US Navy, set up the Wireless Specialty Apparatus Company and manufacture detectors and receivers which are delivered to the U.S. Navy.

In March of 1907 Pickard submits a patent application, which he modifies in November, and for which he is awarded a patent in May of 1909, for a detector using silicon. It is very similar to the one described in his 1906 patent in which silicon having low conductivity is contacted by a good conductor. He uses polished silicon contacted with a brass point, and reports rectification. After polishing, lower pressure is required on the brass point. He states that the rectification action is enhanced by the polishing. But now he knows that his detector is rectifying, i.e., allowing electric current to flow more in one direction than the other, and that the poor conductor or the contact between the poor and good conductors must have some property of rectification.

Silicon and the other low conductivity crystals being used will later be called *semiconductors*. Metal contacts to them can rectify, i.e., pass more electric current for one voltage polarity than the other.

THE AUDION

De Forest continues to use his rectifying vacuum tube detector. In November 1906 he orders a second vacuum tube from Henry McCandless in New York. This time he has a nickel wire positioned between the filament and plate, nearer the cathode. Upon a suggestion from an assistant to

McCandless, they bend the nickel wire into a zig-zag shape. De Forest calls it a *grid*. When he connects a small battery voltage, positive to the grid and negative to the filament, the current to the plate is increased. Negative voltage on the grid reduces the current. A change in this negative voltage can make a large change in the current to the plate. It is a perfect radio wave detector. It responds to a high-frequency alternating-voltage present on an antenna by varying the rectified current through the vacuum tube, battery, and telephone receiver or relay.

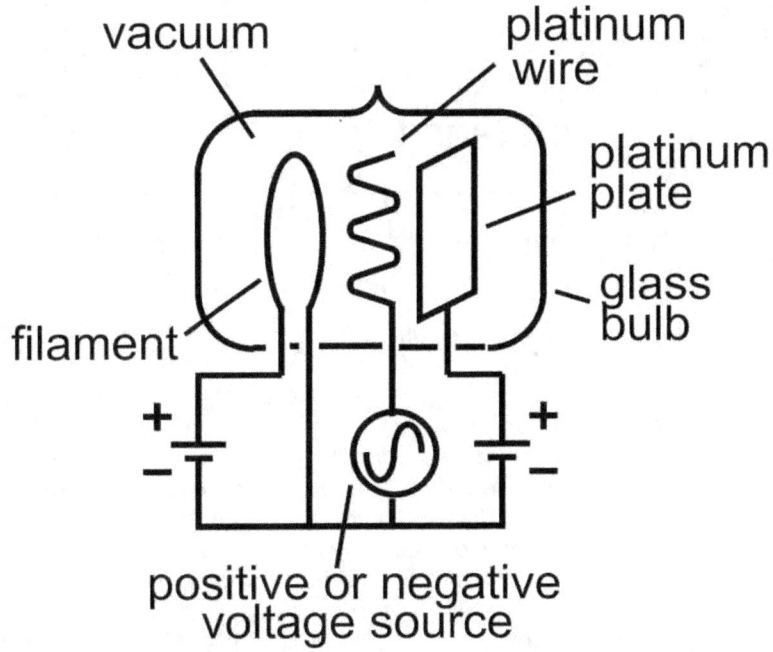

De Forest's Audion having a grid. A varied voltage applied at the grid varies the flow of electrons from the hot filament to the plate

In January of 1907 de Forest files an application for a patent on the vacuum tube with the grid, his Audion. He does not explain the operation of his device, and so is likely unable to explain it. He claims the invention of an "oscillation detector" consisting of a vessel from which air is pumped out and having a filament, plate and, between the filament and plate, a grid, all having electrical wires running out of the bulb. A battery is connected across the filament to heat it. The patent is awarded in February of 1908.

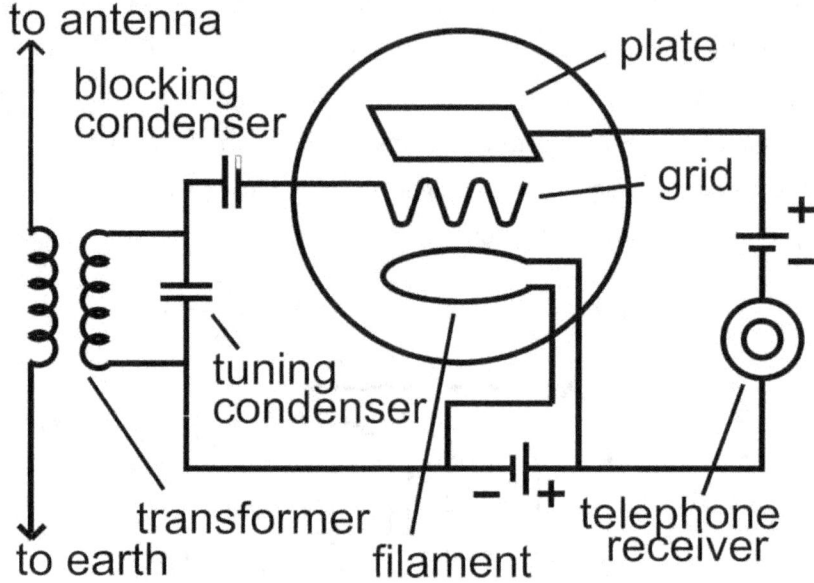

De Forest's radio wave receiver using his Audion with grid. The grid is connected to the negative side of a battery through a winding on an induction coil, now called a transformer, the other winding of which is connected between antenna and earth. A condenser (capacitor) also is connected across the former winding for tuning to a transmitted radio wave. Another condenser blocks battery current that would flow when the grid is positive. The plate is connected to the positive side of another battery, the negative side of which is connected through a telephone receiver back to a circuit common connection.

In 1908, the year Henry Ford introduces his Model T affordable automobile, de Forest is able to send and receive radio messages between buildings three blocks apart and several scientists tune in. He organizes the De Forest Radio Telephone Company to manufacture, market and develop radio. After he narrowly escapes financial failure, the Navy contracts him to equip ships with his radio. In 1908 he goes to Europe, demonstrates transmission from the Eiffel Tower, and sends music 500 miles.

In June 1910 U.S. Post office inspectors find de Forest's company, United Wireless, insolvent in not having the $6.5 million in cash to cover the stocks sold. In 1912, de Forest and company officers will be charged with using the U.S. Mail to defraud investors. He will then offer to sell to the Marconi company the patent rights to his Audion amplifier. In 1913 he will be acquitted of charges but will have to let go of his patent.

John Hammond, a young inventor in his castle-like home in Gloucester, Massachusetts, and Fritz Lowenstein, a physics advisor, correspond in letters that indicate that Hammond is testing a three-electrode tube, the electrodes being filament, plate and grid, as an amplifier, by November of 1911. H.P. Arnold, at American Telephone and Telegraph, believes that a better vacuum would improve the Audion and he tries pumping out more air in laboratory tests. Also in 1911 Irving Langmuir, in the research department of General Electric, has been investigating what goes on inside vacuum tubes when high current flows in a tungsten filament introduced by a Dr. Coolidge. Then Ernst Alexanderson, who built the iron-less alternator for Fessenden, after conferring with Coolidge perfects the Audion and gives it the name that will stick, the vacuum *triode*.

In 1912 the Titanic hits an iceberg and sinks on its maiden voyage. That year Howard Armstrong, a student at Columbia University, experiments with de Forest's Audion vacuum triode and concludes that in it the current consists of electrons only. In an amplifier circuit he connects the plate back to the grid and greatly increases the amplification. This electric circuit loop will become known as a *feedback circuit*, since plate current is fed back to the grid, or a *regenerative circuit*, since a second amplification takes place.

ARMSTRONG

E. Howard Armstrong was born in 1890 in New York City to a father who was a salesman for the Oxford University Press and a mother who had taught in the city schools. They lived in a brownstone on West 29th Street in Manhattan, moved to 97th Street, and finally to a house in Yonkers overlooking the Hudson River. After Howard's father gave him books about inventors and inventions, one chapter in one being on telegraphing without wires, he decided he wanted to become an inventor.

In his room, removed from the family, with a commanding view of the Hudson and the Palisades, he put together radio equipment. His parents supported his purchase of the necessary parts. An uncle introduced him to Charles Underhill, a wireless expert and inventor, whom then Howard visited often for information and answers to questions. He built a powerful receiver and constructed a tall antenna - 125 feet high from ground level.

After passing a bank of entrance exams for admission into Columbia University, Armstrong enrolled in 1909 and commuted from home in order to continue with his experimentation. Armstrong was private and uncommunicative, even in his notebook. He filled his time with the study of radio and held to his vision of a career as an inventor. He completed his coursework, but with a seeming disinterest in the courses, unless they had something to do with electromagnetic waves.

The professor who had the most influence on him was Michael Pupin, a Serbian who immigrated to the U.S. at age 15. Through study and work Pupin had gained a full scholarship from and was admitted to Columbia. He became a teacher of mathematical physics in a new electrical engineering department, and patented inventions in telephone, wireless and x-ray. Pupin had a reverence for pure science but stressed laboratory work over study of theory. He took notice of Armstrong, one of his students in a class on electrical theory.

Now in 1913, five years after de Forest patented his Audion, not de Forest, nor Pupin, nor does anyone else understand fully how it works. Armstrong has concluded from his experiments that the vacuum triode controls the flow of electrons. When used as a radio wave detector he observes oscillations in the electric current flowing to the plate electrode. Then after an idea comes to his mind and he sets up the experiment, he feeds back this voltage to the grid circuit to re-amplify oscillations picked up by an antenna.

Armstrong has put in the plate and battery loop a coil which he places close to a coil in the grid and antenna loop to feed back the amplified signal to the grid. He can hear transmission from Ireland. At high amplification the receiver would begin to hiss. He concludes that the Audion is generating radio waves of its own and that if controlled properly the circuit could be a powerful transmitter as well as receiver of continuous waves.

Not until December does he tell anyone of his discovery, and then only to a friend. He finally asks his father and then an uncle to help him financially to apply for a patent, and they cannot, but his uncle advises him to make a sketch of his circuit and take it to a notary public. With his friend to whom he has finally revealed his invention and who then acts as witness, Armstrong has his drawing notarized.

Early in 1913 Armstrong modifies his amplifier circuit to make it a generator. In March, Columbia professors visit him to see a demonstration and one directs Armstrong to a patent lawyer. But Armstrong has to pass courses and graduate, and he does not contact the lawyer until June to come and see his work. The lawyer tells him to write up a description and file for a patent soon. In the meantime, De Forest and company sell their Audion patent to a lawyer representing the American Telephone and Telegraph in the summer.

It is not until October in 1913 that Armstrong's application is filed. But in the application he does not include the use of the vacuum tube as a generator of continually oscillating electromagnetic waves. On this he applies for a patent in December.

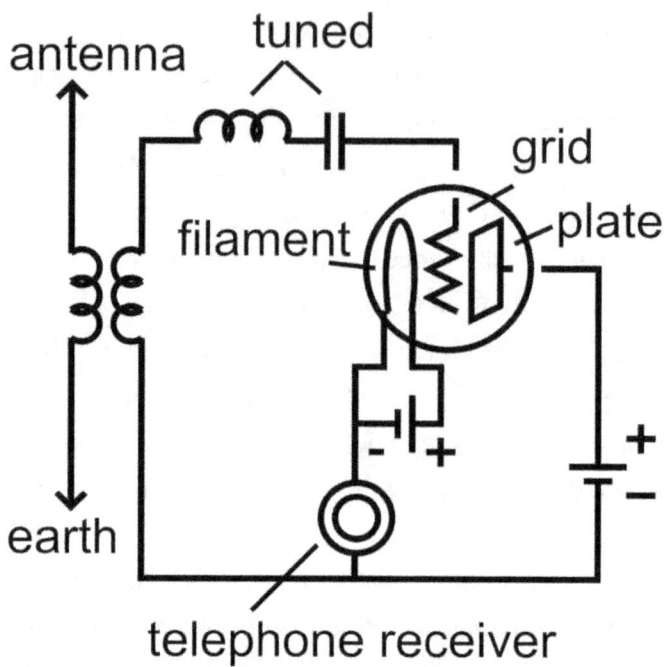

The simplest of Armstrong's EM receivers with feedback. The telephone receiver is in both the grid-transformer-filament and plate-battery-filament circuit loops.

De Forest and another United Wireless engineer are found innocent of fraud in January 1914. In February 1915 in Electrical World de Forest claims he discovered the generating feature of the Audion several years ago, but without filing for a patent.

Armstrong is appointed as instructor to Navy personnel at Columbia. He licenses his regenerative circuit to the Telefunken Company in Germany at the beginning of World War I in 1915 and to the American Marconi Company in 1916. Then in 1917 he heads for Europe as a captain in the U.S. Army Signal Corps in a war against Germany.

CRYSTAL AMPLIFIERS

Certain crystals have been found to be good conductors when one polarity or the other of voltage is applied between them and a metal point contact. The nature of this conduction is not yet understood in 1914 when Eugene Turney in an improved radio wave detector uses a somewhat conducting mineral, preferably galena, and a conducting powder rotated against the mineral crystal to make good contact. In 1920, the copper oxide rectifier is the first to be produced on a large scale. Commercial selenium rectifiers will begin to be made in 1928. Rectifiers employing a metal point contact to somewhat conducting crystals are used as wave detectors in wireless telegraphy, and point contacts to somewhat conducting galena still work the best.

The crystal rectifier is a *solid-state* replacement for the two-electrode vacuum tube, such as Fleming's oscillation valve. Attempts are made at various locations to place a layer containing a metal grid between the electrodes of a selenium or copper sulfide rectifier as is done between filament and plate of vacuum triodes. But the layer is either too thin to insulate the metal grid or so thick as to require too high a voltage.

By 1930, over 40% of American households are reported to have radio, but there is yet no solid-state replacement for the vacuum tube amplifier.

Julius Lilienfeld is awarded a patent that year for a device in which current along rather than through a thin film of a somewhat conductive material is controlled by voltage applied near the thin film, just as current in a vacuum triode is controlled by a grid.

LILIENFELD

Julius Lilienfeld was born in 1882 in what is now Lviv in the Ukraine. He studied at what is now Humboldt University in Germany and received a graduate degree in 1905. He worked with electric discharges at Leipzig University. In his research he applied high electric force to various metals and identified electrons being emitted from them. In the early 1920s he came to the U.S. and in the late 1920s he filed for three patents for radio wave amplifying detectors. He becomes research director for Ergon

Research Laboratories in Malden, Massachusetts and lived with his American wife in Winchester, Massachusetts. They eventually build and move to a house on the Virgin Islands for relief from an allergy, but he travels to the mainland frequently.

He fabricates an amplifier from solid material by depositing two metal electrodes closely spaced onto glass, breaking the glass between the electrodes, sandwiching a metal foil electrode between the two broken edges, depositing copper over the glass and foil edge and then exposing the copper to sulfur to obtain a film of a copper sulfide between the electrodes and over the metal foil line, along the break, which serves as a control electrode.

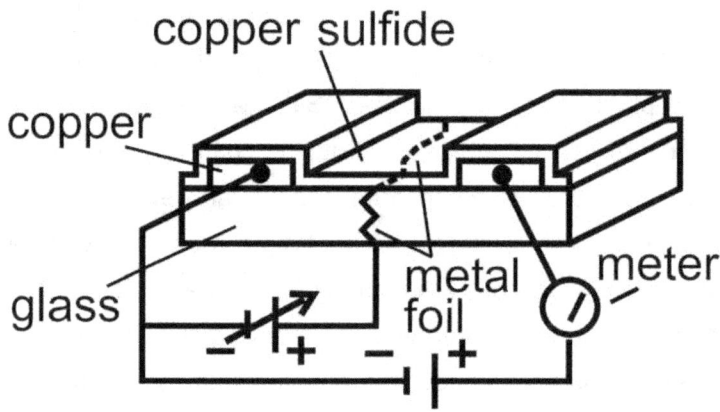

Lilienfeld's solid-state amplifier in which current in copper sulfide film is controlled by a variable voltage applied to a metal foil edge under it

From under the glass he applies a d-c voltage to the metal foil and varies this voltage to vary the conductance of the copper sulfide between the two copper electrodes. He has invented a *solid-state* amplifier. This patent will not result in manufacture but will prevent patents on devices built at Bell Telephone Laboratories years later that perform better but are too similar in construction.

Two years later, Lilienfeld files for a patent for a different version of his thin film amplifier. In this he places a thin insulator between a control electrode and copper sulfide. To make this device a metal plate is oxidized to form a thin insulating metal oxide layer on its surface, and this is coated with copper sulfide. A groove is made in the copper sulfide to make the sulfide very thin in the bottom of this groove. He applies a voltage to the aluminum to control the current flowing from one electrode under the groove in the copper sulfide to the other electrode. Thin copper sulfide in the groove is thereby separated from an aluminum control electrode by thin insulator.

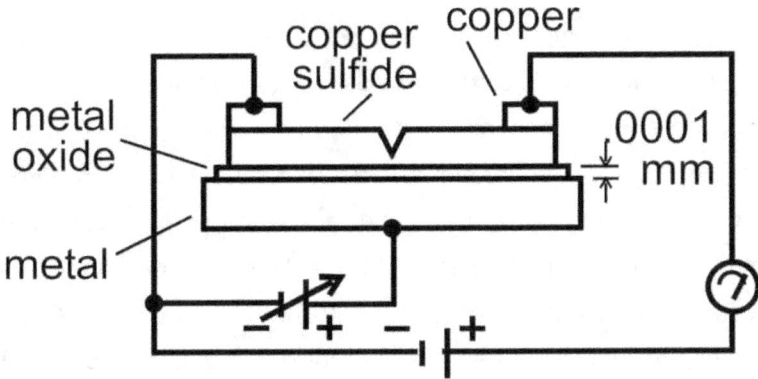

Lilienfeld's amplifier in which current in a thin region of copper sulfide is controlled by voltage applied across a thin insulator

In both of Lilienfeld's devices the current is along a thin layer of a copper-sulfur compound. Attempts are made by Lilienfeld and others to control current, not along, but instead through thin layers by a metal grid between them. Onto a copper sulfide he deposits a magnesium metal film so thin as to have holes through it and over this a second copper sulfide layer. A variable voltage is applied to the porous magnesium film to control the conductance between the copper sulfide layers. He is issued patents on his three devices from 1930 to 1933.

Copper sulfide is not a conductor and is not an insulator. Nor is it now said that copper sulfide is a poor conductor or a poor insulator. It is now classified as a *semiconductor*.

SEMICONDUCTORS

Considerable enlightenment in the identification, characterization and application of semiconducting material is achieved in the 1920s and 1930s, especially in Europe. In 1928 Felix Bloch publishes a theory of movement of electrons through the lattices of atoms of solid crystals. Bloch was born in Zurich, Switzerland, in 1905. He earned an advanced degree at the University of Leipzig under Heisenberg with a thesis on quantum theory. He escapes Europe in 1934 and comes to Stanford University.

Knowledge about semiconductors in the U.S. and Europe is just coming about in the 1930s. Important are an experiment done in England, a suggestion originating in France and a theory proposed in Denmark. Experimental work in laboratories and theoretical work on paper agree and the outcome is a new theory of semiconductor physics. British Alan Wilson after studying for university degrees in Leipzig, Germany in 1931, publishes two papers entitled *The Theory of Electronic Semiconductors*.

Electrons in isolated atoms occupy discrete energy levels. Only one electron can lie at each energy level according to German physicist Werner Heisenberg's *Exclusion Principle*. In a crystal these energy levels spread into energy bands. Between bands there are band gaps. In an insulator these gaps are large, in a semiconductor they are small and under certain conditions electrons can break free into a higher energy band and move through the crystal. In a metal the bands overlap.

It will turn out that germanium can be an ideal semiconductor for solid state amplifiers and then, when problems in the preparation of silicon are solved, silicon will take over. Both germanium and silicon atoms have four electrons in their outer group, *shell, of* electron energies with room for four more electrons to complete a shell of eight electrons. These four electrons are shared between neighbor atoms in a crystal so that each atom has effectively eight electrons in its outer shell.

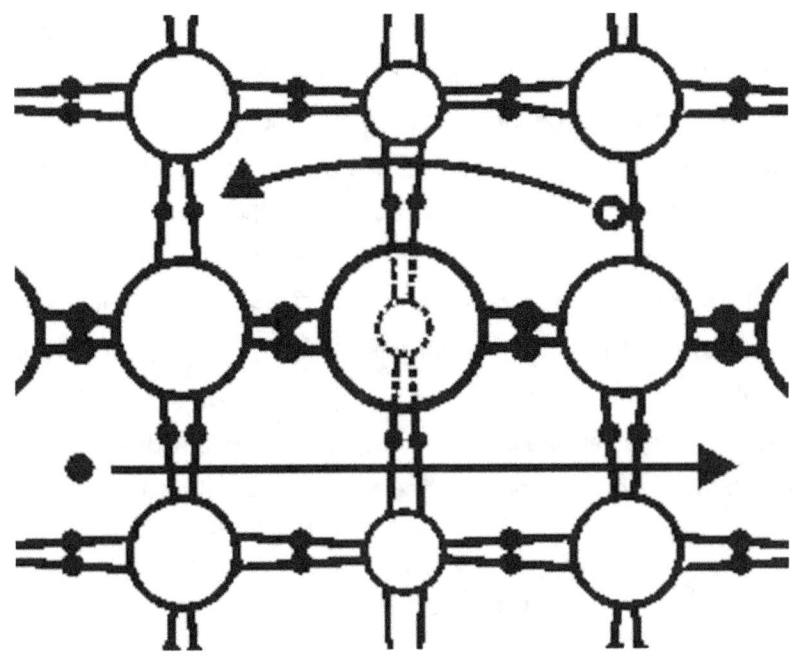

A representation of the 3-D structure of a germanium or silicon crystal showing two atoms (left center and right center) each sharing two electrons with each of four atoms (two in a plane to the front and two in a plane to the rear) and here, one electron having enough energy to pass through the crystal and one hole moving to the left as an electron jumps to a hole to the right.

In the 1930's research is also conducted in various locations to shed light onto the effect, nature and structure of certain pointed contacts to crystal surfaces, specifically how they rectify and conduct. Some crystals are insulators but can be made to conduct by making a metal point contact to them. In 1932 it is proposed that at the surface of a crystal where bonds between atoms are broken there are additional parking places for electrons or additional electrons without parking places.

There are many contributions to understanding made in 1939. Walter Schottky suggests that additional conduction electrons or a deficiency of electrons in the crystal near the surface have energy that differs from the energy of electrons in the point contact metal.(4) This creates a barrier against electron flow between bands unless sufficient voltage of one polarity is applied to lift electrons up and over the barrier.

The crystal radio set becomes important again because now much more is known about crystals. Radio engineers can now build reliable crystal receivers. In the late 30s, war breaks out. The British perfect a radio detection and ranging system to spot German bombers, and they call it *radar*, for RAdio Detection And Ranging. Radar works well using radio waves having a high frequency, but this frequency is so high that vacuum tube rectifiers cannot handle them.

Crystal rectifiers available are inadequate as certain detectors in radar receivers because of low conductance not maintained to a sufficiently high reverse voltage.

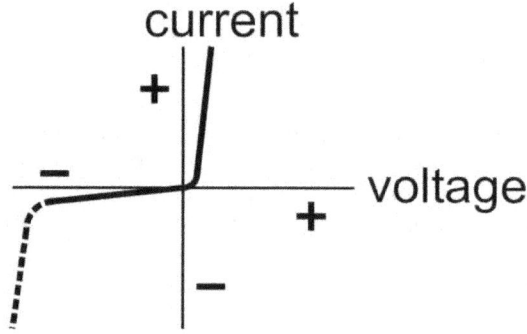

Voltage-current characteristic of a crystal rectifier in which a low conductance is not maintained to high negative voltage

War stimulates search for other materials. Better properties at high frequencies are needed for radar. Fruitful work at Bell Telephone Laboratories, Bell Labs, on germanium rectifiers is done in the 1940s by Jack Scaff and Henry Theurer. At Purdue University, Seymour Benzer discovers the high reverse voltage capability of a rectifier using a point contact to germanium. This fills a need for detectors in radar receivers. In 1943 Wm. Pfann, Jack Scaff and Addison White at Bell Labs file for a patent on a method of shaping metal points for the point-contact rectifier. This involves grinding the point to a desired shape, etching it round and smoothing the tip.

If a small amount of an element with five electrons in the outer electron shell of its atom is added to either germanium or silicon crystals, four of the five electrons are shared in the crystal structure and one electron is available for conduction. The added element is called a *doNor*, its extra electron has a Negative charge and the semiconductor is said to be *N-type*. If the added element atom has only three electrons in its outer shell, it is short one electron to share in the crystal, what is called a *hole* remains, the added element is called an *accePtor*, the hole has a Positive charge and the semiconductor is called *P-type*. Electrons move through the crystal structure playing musical chairs, as electrons move from a crystal atom to a hole in a neighboring atom leaving behind a hole in the atom they left. Holes act as positive charge moving in a direction opposite to that of electrons moving from atom to atom.

P-N JUNCTIONS

At Bell Telephone Laboratories in 1940, Russell Ohl finds a sample of silicon to work better in solar cells than copper oxide or, in use since Pickard's invention, selenium. By melting silicon granules and cooling slowly he and Jack Scaff find the resulting crystal to rectify, due to impurity type difference between two parts of the crystal. Junctions are found between P-type semiconductor and N-type within the crystal.

In 1941, the year Japan bombards Pearl Harbor and the U.S. enters World War II, Ohl applies for two patents not quite one month apart that year, and they are issued after the war. He claims good behavior of silicon as a rectifier at high alternating-current frequencies and with a current for the voltage polarity to block current, *reverse-bias,* much lower than the current for voltage polarity for easy current flow, *forward bias*.

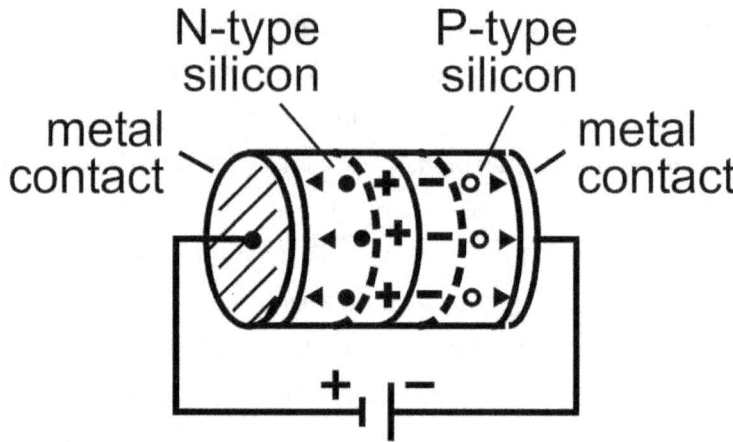

Ohl's silicon P-N junction rectifier with battery connected such that the junction is reverse-biased for low current flow. Electrons (closed circles) and holes (open circles) have flowed away from the junction leaving regions depleted and charged to oppose further current.

Electrons readily flow across the junction from N- to P-region when a voltage is applied across the P-N junction, positive to the P-region and negative to the N-region, *forward bias*. With reverse-bias, negative to the P-region and positive to the N-region, electrons moving away from the junction on the N side deplete electrons in a region near the junction, so leaving positive charge. Electrons move through the crystal from atom to atom toward the junction on the P-side and fill holes by depositing negative charge (depleting holes). These two charged regions create an electric force that blocks further electron flow.

A computer built at the University of Pennsylvania to crack enemy military codes weighs 30 tons and consumes 2000 vacuum tubes each month. A solid-state replacement for the vacuum tube is sorely needed.

At Bell Labs, P-N junctions will be applied in crystal amplifiers as well as crystal rectifiers. Bill Shockley attempts to control conductance in a semiconductor device using an external electric force, similarly to Lilienfeld, but his experiments fail every time.

SHOCKLEY

William Shockley was raised in Palo Alto, California, by a father who was a mining engineer and a mother who became the first female U.S. Deputy Mining Surveyor. William (Bill) graduated from Cal Tech and earned an advanced degree from the Massachusetts Institute of Technology, M.I.T., after doing research on the electrical properties of solids. He went to work for Bell Labs to work for the Nobel-prize winner Davisson, the man who showed that electrons can act like light waves.

After war broke out, Shockley became a research director at Columbia University and then organized a program to train B-29 bomber pilots on the use of new radar bomb sights. He wrote a report on probable casualties among soldiers and civilians if Japan were invaded and this report likely influenced the decision to drop the two atomic bombs.

Back at Bell Labs in 1945, a research group is charged with finding a solid-state replacement for the vacuum triode and for the mechanical relay. The vacuum tube filament must be heated and so wastes power. The relay operates at only mechanical speed and is only either closed or open, so there are no intermediate states. It is believed that a solid-state device could be fast, cheap and reliable. Shockley is made the group leader at Bell Labs, which is now a creative environment at Murray Hill in New Jersey.

Walter Brattain, an experimental physicist, and John Bardeen, a theoretical physicist, are assigned to Shockley's group. The first patents for which they file are rejected because of similarity to the earlier work of Lilienfeld. But Brattain in February 1948 files for a patent on a semiconductor device in which current along a semiconductor layer is controlled by applying an electric voltage to a drop of liquid on the semiconductor surface. Electric control force is transmitted through the liquid with little input current required. Water works well as the liquid. Varying this electric force varies the output load current.

BRATTAIN

Walter Brattain was born in 1902 to American parents in China where his father was a teacher. After coming to the U.S. in 1903 they lived near Spokane, Washington. Walter attended Whitman College and majored in physics and mathematics. He earned degrees there and at the Universities of Oregon and Minnesota. He joined Bell Labs in 1929. He is in the solid-state research group created in 1945 to work on semiconductor amplifiers under Shockley.

After the failures to obtain amplification by applying electric force externally to a semiconductor as proposed by Shockley, John Bardeen suggests that these failures are due to a layer of electrons on the surface of the semiconductor, changing the semiconductor type at the surface. There are extra electrons on the surface and these are adequate to shield the semiconductor from an external electric force.

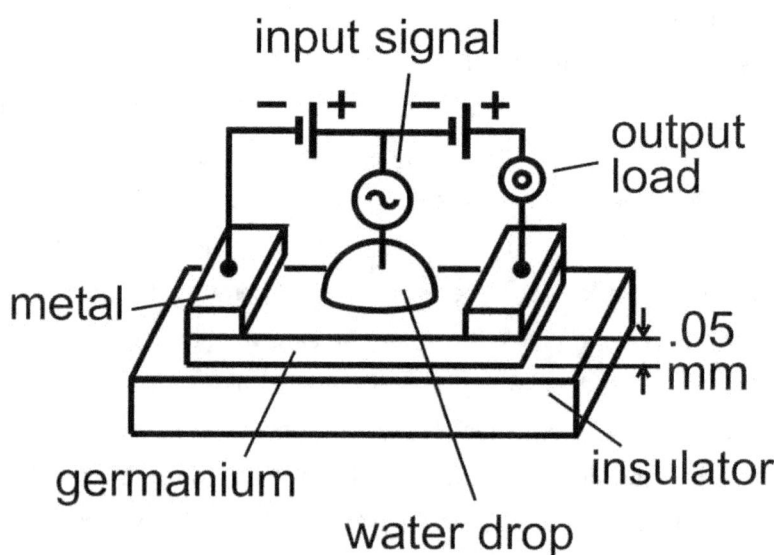

Brattain's semiconductor amplifier in which current along a thin germanium crystal is controlled by voltage applied to a water drop on the germanium surface.

BARDEEN

John Bardeen was born in Wisconsin in 1908, the son of the dean of the medical school at the University of Wisconsin. He graduated from high school at age 15 and graduated from the university with an advanced degree in electrical engineering in 1929. After working a short time for Gulf Research Labs, he began graduate study at Princeton University in mathematics and physics. Before graduating he worked at Harvard University for three years on cohesion and conduction in metals and finished work for an advanced degree in mathematical physics. He is hired by Bell Labs in 1945.

On the same day that Brattain files for his patent involving the controlling of current along a thin semiconductor layer by applying a control voltage to a drop of water, Bardeen files for a patent on controlling the current along the thin N-type surface on a P-type germanium block. A small point contact is made to the surface and a large area contact to the underside of the block. Voltage is applied between the contacts such as to slightly reverse-bias the P-N junction under the point contact. Electrons move in toward the point contact. This current is controlled by a voltage applied to a ring around the point contact, but insulated from the germanium surface.

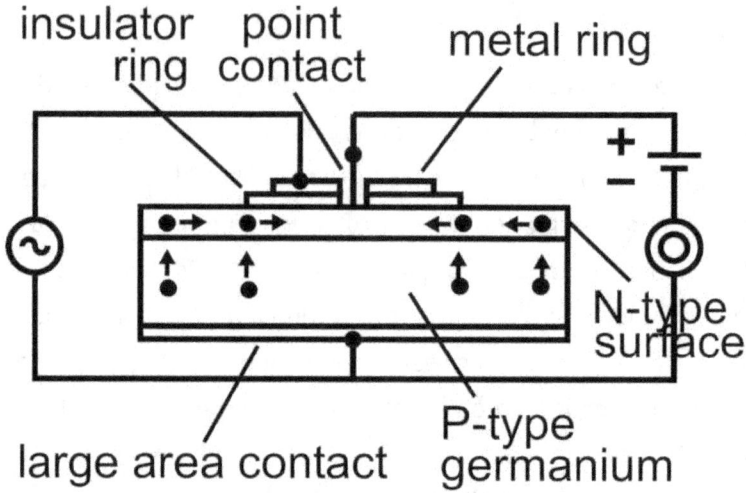

Two-dimensional representation of Bardeen's semiconductor amplifier with a metal ring controlling electron current flowing in toward a point contact from out along a reverse-biased P-N junction.

THE POINT-CONTACT TRANSISTOR

By 1946, two critical decisions have been made at Bell Labs, these being that the semiconductor amplifier would make use of electric force, as did Brattain's and Bardeen's amplifiers and the semiconductors to be investigated would be germanium and silicon. Bardeen completes a theory on semiconductor surfaces. Brattain experiments with point contacts to semiconductor surfaces. He and Bardeen try to confirm Bardeen's theory of electrons and electric force near point contacts. They have success with an amplifier using two point contacts on a germanium crystal and they demonstrate their device at Bell Labs in December of 1947.

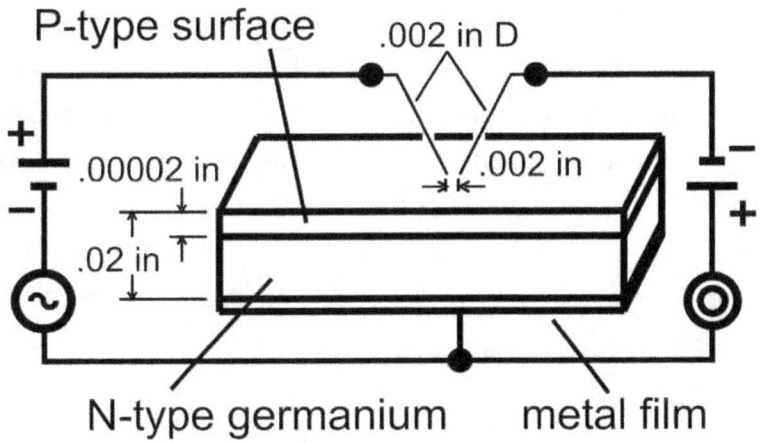

Bardeen and Brattain's semiconductor amplifier having two point contacts to a P-type germanium surface of a tiny N-type germanium semiconductor crystal

To avoid conflict with Lilienfeld's patents, the patent attorneys at Bell Labs in June 1948 file for a patent on Bardeen and Brattain's device as a Three-Electrode Circuit Element Utilizing Semiconductive Materials. Also in June the U.S. Defense Department in a meeting in New York at Bell Labs is warned, before it is announced to the American press, about the importance of this new device for possible military application. Bell Labs calls the device a *transistor*, a shortening of two words - transfer resistor, in which the transfer of signal is controlled through the device's *resistance* (the inverse of conductance: R = 1/G where G is the commonly used symbol for conductance).

So mid-20th century the transistor is born and given a name. But it will evolve into a very different device by further invention.

In their patent Bardeen and Brattain explain the main principles of operation: The invention uses a semiconductor crystal on top of which two rectifying point contacts are made, one called an *emitter* and the other a *collector*, and a high-conductance contact called the *base* is made to the bottom. A forward-bias voltage, the polarity for high conductance, is connected between the emitter and the base. If the semiconductor is N-type, there is a P-type surface layer on the N-type semiconductor. Forward bias is positive to the P-type layer. Electrons in the P-type crystal flow out the emitter, so that the emitter, in effect, is emitting holes into the P-type layer. Holes flow from emitter to collector, to which a large negative reverse-bias voltage is connected, which injects electrons, i.e., attracts the holes and collects them. Only a small voltage change is required at the emitter to change the hole current to the collector. This can allow a large change in the reverse-bias voltage between collector and base, and thus produce a large power amplification, power *gain*.

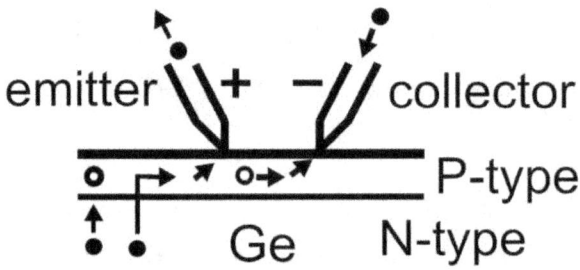

A two-dimensional representation of the flow of electrons and holes in the point-contact transistor. Electrons flowing from the emitter, at forward-bias voltage, leave holes which are attracted to the collector, at large negative reverse-bias voltage.

More detail will be given in Coblenz and Owen, Transistors: Theory and Operation, 1955. On an N-type germanium pellet 20 mils (thousandths of an inch) thick and 50 mils across, Bardeen and Brattain place metal point contacts, called cat whiskers, only 2 mils apart and bend them for a light pressure of point contact onto the germanium surface. A soldered high-conductance base connection is made to the bottom of the pellet. A positive electric voltage is applied to the emitter point contact which pulls electrons from atoms in P-type semiconductor to inject holes which move toward the collector where they are filled with electrons available at the collector. Since the emitter and collector point contacts are closely spaced, many holes reach the collector. The emitter at low forward-bias voltage emits holes into the semiconductor which are collected by the collector at high reverse-bias voltage.

If P-type germanium crystal were used, an N-type surface layer would form, negative voltage would be applied to an emitter cat whisker which would emit electrons into the germanium, many of which would be collected by the collector cat whisker to which a large positive reverse-bias voltage is applied.

THE JUNCTION TRANSISTOR

In January 1948, learning from Brattain and Bardeen's discovery, Shockley completes a theory on junctions between the two types of semiconductor. He records in his notebook an idea for an amplifier employing three semiconductor regions in one crystal of germanium, with semiconductor of one type sandwiched between semiconductors of the other type. He proposes that if electrons could be emitted at a varied low voltage into the N-side of a forward-biased P-N junction, many could cross a thin P-region and a reverse-biased second P-N junction and be collected at a large voltage which then varies for amplification of voltage.

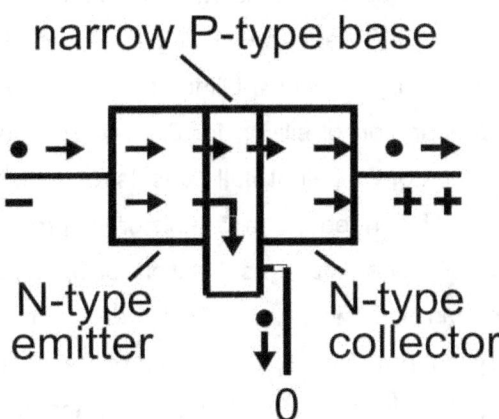

Shockley's junction transistor design in which many electrons emitted across a P-N junction at low forward-bias voltage cross a second P-N junction at high reverse-bias voltage and are attracted to the positive collector. The sandwiched layer must be very thin and it, for connection to it, must extend beyond a layer over it.

Shockley's name is not included in Bardeen and Brattain's application for a patent on their point contact transistor. Shockley, not to be upstaged, files for his own patent two weeks later on various versions of his "sandwich" amplifier.

Shockley does not give details of fabrication of his three layer device, since one has not yet been fabricated, but does explain the theoretical physics of semiconductor junctions. He secretly develops on paper his transistor he now calls a junction transistor, employing two junctions between N-type and P-type semiconductor in three layers. He steers his group away from basic research to projects having to do with his transistor. He micromanages and tries to direct efforts toward his own ideas.

The research phase at Bell Labs is ending. Bardeen and Brattain move to another group. Neither then have much to do with transistor development. Bardeen applies to the University of Illinois and Bell Labs tries to keep him, but in April of 1951 he leaves. They will meet again with Shockley when they are awarded a Nobel Prize for the invention of the point contact transistor.

In 1948, the year Columbia Records introduces the long-playing vinyl disc record for "hi-fi" sound, G.K. Teal and J.B. Little succeed in growing a single crystal of germanium by slowly pulling a seed crystal from molten germanium. For preparation of silicon for Shockley's transistor the work of Scaff and Theuerer and of Ohl, all at Bell Labs, is recognized.

Bell Labs learns of a need for a transistor in proximity fuses for the Korean War and Shockley encourages the work of Sparks. Teal and Sparks at Bell Labs have learned how to grow N-P-N germanium crystals by pulling an N-type seed crystal from molten P-type and obtain P-layers only 1-2 mils thick. In April 1951 Shockley's junction transistor is announced at a Bell Labs press conference.

In an N-P-N transistor, narrow P-type semiconductor is sandwiched between N-type semiconductors. In a P-N-P transistor, narrow N-type semiconductor is sandwiched between P-type. The structure of these transistors is similar to that of others who have attempted a grid, but the principles of operation are very similar to those of the point contact transistor. The junction transistor is the result of calculations and understanding instead of experiment and discovery and will prove to be adaptable to manufacture in large quantity.

Both transistors can be used in an amplifier circuit in which the input alternating-current signal is connected to the base and a low-conductance load is connected to the collector. A varying voltage applied to the base varies the electron injection into the emitter. Many electrons coast on

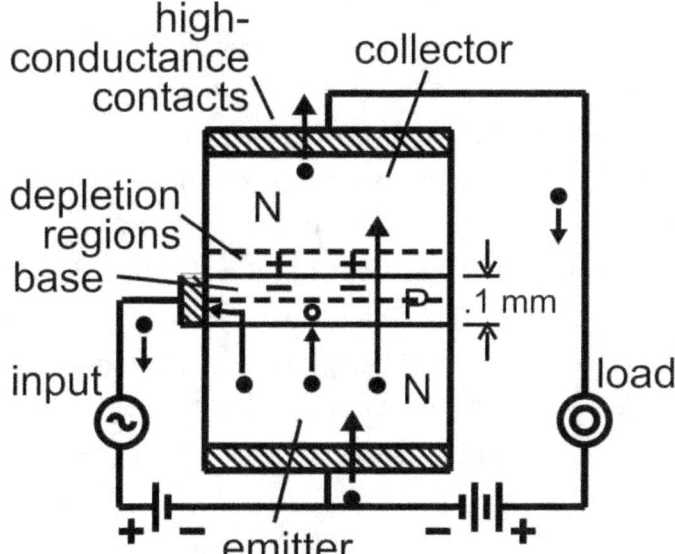

through the base and are driven across a reverse-biased collector P-N junction by the charge left by depletion of electrons and holes.

In the N-P-N junction transistor some electrons entering the lower N-region (emitter) at forward-bias voltage exit the P-type base or fill holes in the base but most are driven through a depletion region by charge existing there due to reverse-bias voltage.

There are advantages in using either transistor in an amplifier circuit in which the transistor emitter is common to input and output, i.e., a circuit in which the input current is the small base current and the output current is the larger collector current. The input voltage applied between base and emitter is small and the output voltage taken between collector and emitter can be large. High current gain and high voltage gain result in high power gain.

Shockley publishes his book titled *Electrons and Holes in Semiconductors* in 1950. He is awarded a patent for the junction transistor in September 1951.

In 1951 Shockley, just after his work on transistor operation involving both electrons and holes is published, in a patent application proposes for a very different transistor, unipolar operation in what is a type of field-effect transistor. When current is by electrons in N-type semiconductor, electrons will move further, there being few holes into which to fall, *recombine*. This increases switching speed and amplifying frequency. In one embodiment of his invention in a P-N-P structure the electron current flows through an N-type region parallel to, and not through, P-type regions on either side. Reverse voltage is applied between N-type and P-type regions, sweeping electrons and holes away from the junctions and creating an electric force that blocks electron flow across this junction. This constricts the path for electrons in and reduces the conductance of the N-region. The conductance varies as the reverse-bias voltage applied to the P-region is varied.

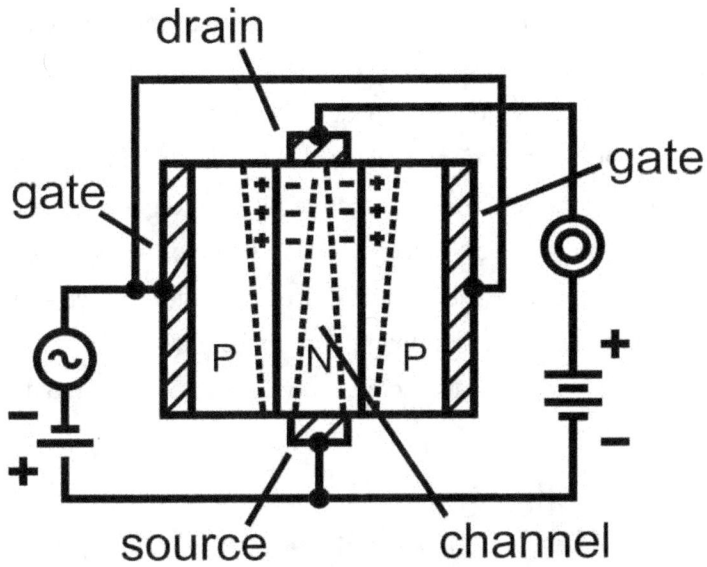

Shockley's junction-gate field-effect transistor. Reverse-bias voltage is applied between the gates and source and a larger reverse-bias voltage is applied between the drain and source. The regions depleted of electrons and holes and left with charge constrict the channel, more so at the drain end, due to a larger gate-to-drain voltage.

A battery is connected between ends of the N-region and electron flow is from one end, which Shockley calls the source, to the other end, the drain, where the path, called the channel, is more constricted, pinched, by the battery voltage. If this reverse-bias voltage is increased enough to close the channel, pinch-off is said to occur and channel current increases little with further increase in channel voltage. But a small change in the gate voltage produces a change in electric force, i.e., electric field which is electric force per unit electric charge, at the semiconductor surface. This results in a large change in the current flowing in the channel and a large change in voltage across the load, e.g., a speaker or headphones.

Another version of this insulated-gate field-effect transistor will become very important, later, in integrated circuits. But improvement continues on bipolar junction transistors, in which both electrons and holes play a role in all.

In 1952 J.E. Saby of General Electric announces the development of an alloy junction transistor. Dots of P-type impurity are deposited onto both sides of thin slices of N-germanium and the germanium slice is baked to diffuse in the impurity to form an emitter and a collector on opposite sides of an N-type transistor base. The slice is then diced into many transistors. The frequency range of operation is improved by making the base very narrow. Base width of 10 microns (thousandths of a millimeter) is achieved and the transistors operate up to an alternating-current frequency of millions of cycles per second. It is readily manufactured and will become a mainstay of the industry.

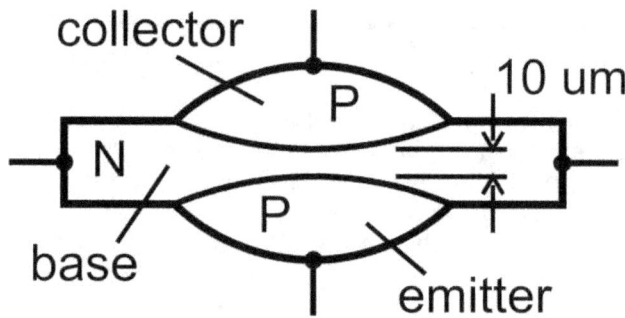

Alloy junction transistor (cross-sectional view). Impurity is diffused into the N-type germanium slice from P-type dots.

Texas Instruments supplies transistors for 100,000 transistor radios sold by Regency that year.

Research is ongoing in silicon, which has better characteristics for transistors than germanium. In 1954 Fuller and Pearson diffuse impurity into N-type silicon to form a thin P-type layer for Bell Labs' P-N junction diode solar cells. Morris Tanenbaum diffuses two impurities into N-type silicon, the first going slightly deeper for a P-type base and the second converting P-type silicon back to N-type, leaving a 0.1-mil (2.5 micron) thick base for an N-P-N junction transistor

Shockley leaves Bell Labs, in 1956, and moves to Mountain View, California, to be closer to an ailing mother, it is reported, and to set up a division of Beckman Instruments, that he calls Shockley Semiconductor

Laboratories, in Palo Alto, to manufacture transistors. His company is the first working on silicon semiconductor devices in the region that will become known as Silicon Valley. And so it is Shockley who brings the silicon to Silicon Valley.

Shockley hires some of the brightest engineers in the U.S. But it seems Shockley has a superiority complex. He appreciates little his employees' opinions and preferences. When he stops research on silicon transistors, eight of his best men leave his company and set up a division of Fairchild, which they call Fairchild Semiconductor. In 1963 he will be eased out of his own company and will join the faculty at Stanford University, ending his industrial career at age 53. Shockley will not supervise or contribute directly to the development of the *planar process* that will lead to the *integrated circuit*.

CHAPTER 3 - INNOVATION

THE PLANAR PROCESS

Research in semiconductors has yielded three types of transistors - point-contact, junction and field-effect. Application of these transistors, especially the field-effect transistor, will make possible smaller, faster and less expensive electronic circuits, and by the end of the 20th century, the smart phone.

In 1957, the year the Soviet Union launches the world's first artificial satellite, Fairchild Semiconductor occupies a rented building in Palo Alto to manufacture transistors. The company's assets are not much more than the eight men who left Shockley plus five who join them. At Fairchild they continue their work which they had begun at Shockley Semiconductor on silicon transistors.

Their best transistor is the double-diffused. They use heat and pressure to diffuse an impurity into silicon of one semiconductor type to create a region of the other semiconductor type. Into the latter type they then drive in the first impurity almost as deep for an emitter region leaving a base region, very thin (narrow). This results in a transistor capable of switching at high speed and amplification at high frequency.

Not many of the transistors pass the necessary tests at the end of the fabrication process. This double-diffused transistor's P-N junctions are exposed at their edges. One huge problem is dust left on the surfaces of the transistor at these edges causing electric current to detour around the junctions.

Fairchild's research physicist, Jean Hoerni, has been studying this problem, from the time the manufacturing process was being set up there. He comes up with the idea of intentionally oxidizing the silicon surface to create a silicon oxide insulator there, even before the impurity diffusion is carried out. Silicon dioxide will play a big role in the development of the integrated circuit.

HOERNI

Jean Hoerni was born in Geneva, Switzerland in 1924. In 1952 he moved to the U.S. to work at the California Institute of Technology. There he met Shockley, who recruited him to work at Shockley Semiconductor. Jean was one of the eight who left Shockley's company and set up Fairchild Semiconductor.

Hoerni looks for a way to insulate the silicon from the destructive tiny stray dust particles left inside the finished packages. Researchers at Bell Labs have worked on the problem without finding a solution. Hoerni still believes there is a solution, and his solution is to oxidize the entire surface of the silicon slice, *wafer*.

To diffuse impurity into the silicon, he reasons that the oxide can be removed from a portion of the top of wafer to create a window and an impurity driven in with heat and pressure through this window to create a transistor base inside a transistor collector. The surface can be again oxidized, and inside this transistor base, the oxide can be removed from a smaller area through which to drive in another impurity to create an emitter region. The entire surface is then oxidized and windows etched in the oxide for making electrical connections to the three regions.

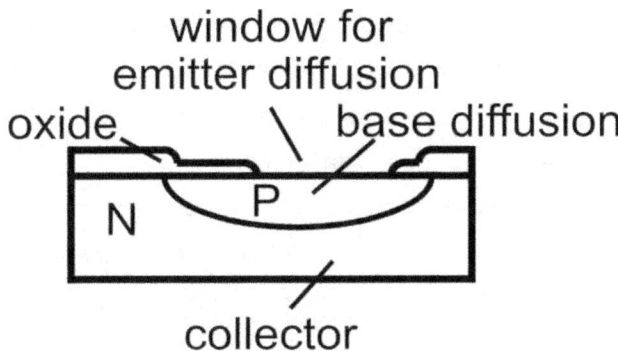

Hoerni's partially completed planar N-P-N transistor. The emitter diffusion into the base is to be slightly less deep than the base diffusion into the collector. This emitter region is off center in the base to allow room for wire connection to the base. The P-N junctions on both sides extend to the top surface of the silicon wafer, which is covered by oxide.

Hoerni works nights, when manufacturing equipment is available to him, to fabricate transistors with this process, which he calls the *planar process*. The transistor is then a planar transistor, to which electrical connections to the collector, base and collector are made on its top surface, a flat plane, after the silicon wafer has been diced into chips.

The final test of the manufactured transistor is to tap it to move dust around inside the transistor can. It will be said that Hoerni whacked the can with a hammer and the transistor still passed the test.

Hoerni files for a patent he titles, Method of Manufacturing Semiconductor Devices, in May of 1959, in which he claims the protection of the silicon by silicon oxide, adding or removing oxide in defined places on the silicon wafer and controlling the diffusion of impurities to change semiconductor type. Short circuits around the junctions are prevented. All the while, the wafer is protected by oxide.

The manufacturing process steps are essentially (1) forming the wafer, (2) coating it with oxide, (3) etching away portions of the oxide to form a mask, (4) driving one type of impurity into the silicon through the openings in the mask with heat and pressure to alloy the impurity with the silicon and (4) recoating the openings with oxide. This is repeated for another type impurity. For electrical connections, the oxide is etched a final time, (5) gold is plated through the openings in the oxide and the wafer heated to alloy the gold with the silicon to make a high-conductance contact.

The patent is awarded to Hoerni in November of 1962. But in the meantime, an electrical engineer, Jack Kilby, at Texas Instruments has come up with a whole circuit in one crystal, an *integrated circuit*, at that company.

THE INTEGRATED CIRCUIT

KILBY

Jack Kilby was born in 1923 in Missouri and attended school in Great Bend. He went to college at the University of Illinois and completed the requirements for a master's degree in electrical engineering at the University of Wisconsin while working for Centralab in Milwaukee. Upon graduation he goes to work for Texas Instruments, T.I., in Dallas, Texas, in 1957 when at the time the company is manufacturing and marketing germanium and some silicon transistors, and has money to spend on research.

Effort has been spent on reducing the size of components and packing them close together, but the methods require extra operations and many materials and tests. The connections between circuits use some of the area on the semiconductor wafer. Reduction in size has nearly reached a limit. His first summer at T.I., Kilby has yet not earned vacation time, so he works on this problem, which is called the *Tyranny of Numbers* problem, and concludes that a solution might be the making of the components of an electronic circuit together in one semiconductor crystal.

In July 1958 Kilby records his idea in his laboratory notebook. In September he makes various semiconductor components using germanium for the semiconductor. Then he is ready to make a whole electronic circuit out of one crystal of germanium. His supervisor gives him approval to use people and equipment to carry out the task and assigns an oscillator (a-c generator) as the electronic circuit to integrate. Kilby glues the germanium crystal to a support block, connects the components of his circuits with gold wire and runs out larger wire to a supply of direct current for power and to an oscilloscope for observation of the circuit's performance. In September with others looking on, he turns on the direct current power to make appear on the screen of the oscilloscope an alternating electric voltage.

Kilby's idea is presented to the U.S. Air Force, which had been encouraging development of a new way to build electronic equipment. The Air Force accepts the idea and grants Texas Instruments a contract.

The managers at Texas Instruments are worried that another company is about to patent an integrated circuit, but it being RCA, not Fairchild. For the patent application, they need a picture of the invention, but the only model they have is Kilby's germanium device, a crudely constructed demo version, with gold wires connecting the components of the *integrated circuit*. To hurry along the patent application, they use a picture of this demo device, but write that changes and modifications are possible which do not depart from the concepts. Finally, they add that insulator may be formed onto the semiconductor wafer and etched and gold deposited to make the electrical connections.

In February of 1959 Kilby's application for a patent is received at the U.S. Patent Office. In it is described a circuit fabricated from a tiny semiconductor bar in which planar transistors and capacitors, each made from P-N junctions, are shaped, i.e., have their sides etched away to isolate them from other components. Resistors, which slow electric current and have desired conductances, are semiconductor regions isolated from other components by reverse-biased P-N junctions. The capacitor capacitances and the resistor conductances are set by their areas and the amount of impurity driven into these areas. The resistor conductances also depend on the depth of the impurity diffusions.

Kilby describes two possible circuits formed in a P-type germanium or silicon crystal in which are resistors in P-type semiconductor and capacitors and transistors in P-type regions into which N-type impurity is diffused for P-N junctions. Transistor bases are N-type and have rectifying metal contacts for transistor emitters. Transistor areas are reduced by etching away their perimeters. Conductance in P-type semiconductor is set for resistors connecting other components. Capacitance in P-N junctions on top of resistors is used for capacitors. Interconnections are made with gold-plated wires.

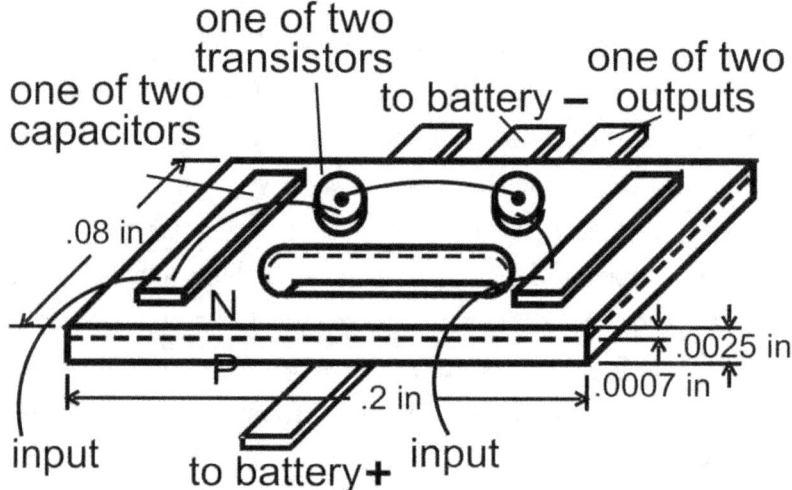

The first integrated circuit. Kilby's semiconductor generator circuit integrates two transistors, two capacitors and eight resistors. The slot in the semiconductor chip defines the shapes and therefore the conductances of the semiconductor resistors. The five curving wires connect together aluminum transistor emitters on N-type-bases, connect the N-type bases to other components and provide input connections.

Oxide insulator covers the semiconductor except where connections are to be made. Some connections to the transistor bases and emitters and other points in the circuit are made with gold plated wire. But the patent states that connections may be made in other ways, such as with a deposited metal layer, or as written in the patent, gold LAID DOWN ON. These last three words will be important in future patent hearings.

The number of fabrication steps is small, the devices are very compact, inexpensive and reliable. Spaces between components are extremely small and component counts can be 200,000 per square inch, as compared with 3500, the highest density reached with discrete (separate) components. However, Kilby's patent for which he applied in 1959 is not granted until 1964.

In the meantime, Robert Noyce, one of the eight who left Shockley and set up Fairchild Semiconductor, in 1959 has filled four pages of his laboratory notebook with a description of a possible integrated circuit, fabricated from silicon coated with silicon oxide using Hoerni's method. He has written that in many applications it would be desirable to make multiple devices in a single silicon body.

NOYCE

Robert Noyce grew up in Grinell, Iowa. At Grinell College he studied physics, and he took time to participate in diving, music and drama. It is reported that one time at Grinell, he stole a pig from the college farm and roasted it for his classmates. As a consequence he was suspended, but only for one semester.

One of Noyce's physics professors showed his class two of the very first transistors to come out of Bell Labs, and Robert was fascinated. He went on to M.I.T. in Massachusetts, received an advanced degree in 1953, and went to work for the Philco Corporation. He moved to California in 1956 to work for Shockley Semiconductor and left Shockley in 1957 as one of the "mutinous" eight. At Fairchild Semiconductor he was elected CEO by his fellow workers.

After Hoerni's success in making their double-diffused silicon transistor profitable, Noyce studies the possibilities of making a whole circuit the same way. In January of 1959 in the four pages of his notebook he writes that if multiple transistors, resistors and capacitors are made in one silicon crystal, it would be possible to make interconnections between them with metal deposited during the chip fabrication process and thereby reduce size and cost. He sketches a simple calculator circuit to add two numbers, the circuit made in one body of semiconductor. Noyce has what is called the *Monolithic Idea*, an idea for an electronic circuit manufactured by employing a single body of semiconductor on the surface of which are done the deposits, etchings and diffusions in steps for a whole circuit, including interconnections.

By the Spring of that year, rumors of Kilby's work at Texas Instruments have reached someone at Fairchild Semiconductor. Noyce has first worked out, applying Hoerni's results, the possibility of interconnections between transistors and then considered the possibility of integrating transistors into one silicon body. They decide to file for a patent, but they know that Texas Instruments has already filed. So they need a "legal shield" that would differentiate Noyce's version from Kilby's. Noyce's ideas have begun with the component interconnections and from there he considers integration of whole circuits. He can supply a description and a drawing of a semiconductor body with interconnections built in. They title their application Semiconductor Device-and-Lead Structure and make its first objective the improvement of a structure for making electrical connections to various semiconductor regions.

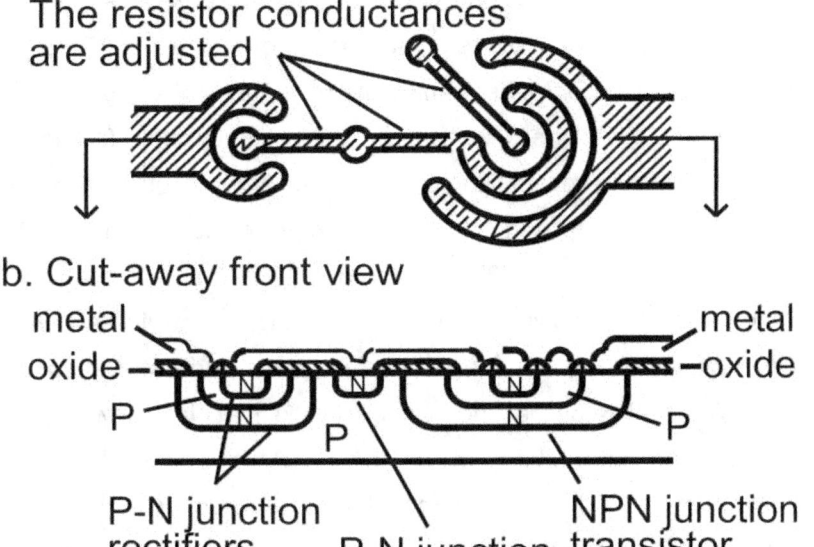

Noyce's Device and Lead integrated circuit using metal conductor deposited onto the surface of a semiconductor crystal to connect together the components of a radio wave receiver. Two rectifiers (P-N junctions) detect voice on the peaks of high-frequency antenna alternating-voltage, a capacitor (P-N junction) and two resistors (low-conductance conductors) filter out the high frequencies and a junction transistor (N-P-N) amplifies the voice voltage. a: top views of conductors; b: cut-away front view of P-N junctions.

Fairchild adds three pages of circuits that can be integrated and files the application in July 1959, two months after Kilby's application. It goes to a patent office examiner who moves it along so quickly that the patent is granted before Kilby's. Then at Texas Instruments, a patent lawyer files papers for an Interference Proceeding before the Board of Patent Interferences to determine which inventor had the idea first. Kilby has entered his idea into his notebook in July 1958. Noyce has done the same in January 1959.

The Fairchild patent application describes a semiconductor body having dish shaped junctions formed by two diffusions of impurities. These junctions, where they extend to the surface of the chip, are covered by an insulator that is an oxide of the semiconductor. Connections within and to the circuit are formed as STRIPS of metal that extend over and are ADHERENT TO the oxide layer. Noyce will be granted the patent in April of 1961.

But Kilby is the winner decided by the Board of Patent Interferences. A new lawyer at Fairchild focuses onto Kilby's interconnection with wires or gold "laid down" as opposed to Noyce's interconnection with material "adherent to" the oxide. In 1967 the Board will not be impressed by the difference between "laid down on" and "adherent to" and will rule in Kilby's favor. Fairchild will appeal and the Court of Customs and Patent Appeals in 1969 will decide that Kilby had not demonstrated that "laid down on" had acquired a meaning in the art and will reverse the Board's decision.

Texas Instruments, T.I., will then appeal to the Supreme Court, which will deny the request to review the opinion. Noyce and Fairchild win, but by this time, the market for integrated circuits will have exploded and executives from T.I., Fairchild and about a dozen other firms will hold a meeting and cut a deal. T.I. and Fairchild will agree to grant licenses to each other for integrated circuit production. The agreement will provide T.I. and Fairchild with more than $100 million in royalties over the following years.

After ten years of legal battle, court fees and countless documents written and studied by numerous lawyers and engineers, the Noyce/Fairchild final victory will mean little. Both Noyce, with the help of talented engineers and scientists who worked with him, and Kilby, with the support of a large company by whom he was employed, will be considered co-inventors of the integrated circuit.

THE MOSFET

Work on an insulated-gate field-effect transistor has been continuing at Bell Labs. It is a planar device and is well suited to be incorporated into single-crystal integrated circuits. However, it relies on silicon oxide on a silicon surface, which is found to be not very stable electrically and this is a problem. Research on it is progressing, particularly in laboratories at Fairchild Semiconductor and also at Bell Labs and IBM.

In 1958, M.M. (John) Atalla and his group at Bell Labs find that by carefully cleaning the surface of a silicon wafer and creating a thin silicon oxide layer onto it, reduced in number are electron *traps* at crystal atoms, where conduction electrons fall into holes and are removed from the action. Oxygen in the oxide at the silicon surface bonds with silicon atoms not having other silicon atoms with which to bond. Attalla reports in a paper in 1958 that his group has grown a thin layer of silicon dioxide on the outer surface of silicon in silicon rectifiers and this reduces the reverse-bias current, by factors of 10 to 100. Electrical noise caused by trapping and detrapping of electrons at the surface that has been a problem at an alternating-current frequency of 10,000 hertz is reduced so much that it is not a problem even at higher frequencies.

ATALLA

John Atalla was born in Egypt in 1924. He came to the U.S. to do graduate study in mechanical engineering and went on for an advanced degree at Purdue University awarded him in 1949. He went to work for Bell Labs doing research on, among other topics, silicon oxide used as a protective layer on silicon devices.

In 1959, business is booming. Texas Instruments sells $90 million worth of diffused silicon transistors and diodes, Transitron in Wakefield, Massachusetts and General Electric $30 million each, and Fairchild Semiconductor $7 million.

In May 1960, Dawon Khang in Attalla's group files for a patent for a device they call an electric-field-controlled semiconductor device. This invention, or design, of an insulated-gate FET is like the junction transistor in one respect in that it employs two separated P-N junctions, one across which forward-bias voltage is applied and the other across which reverse-bias voltage is applied, but current across these junctions is controlled not by a base current, but by voltage applied across an insulating layer.

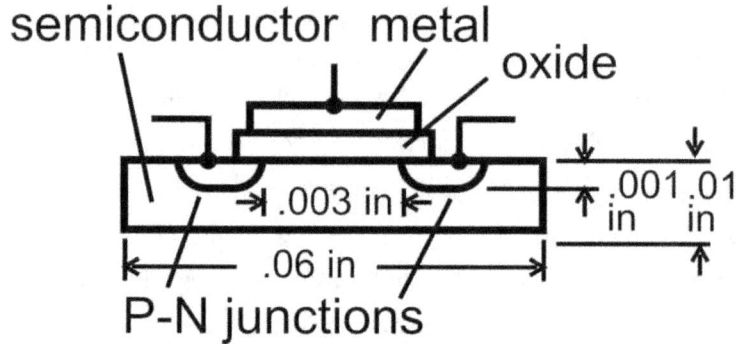

Khang's electric-force-controlled semiconductor device. The voltage to one P-N junction is that for forward-bias for easy flow of electrons and voltage to the other junction is that for reverse-bias. The conductance of the semiconductor between junctions is controlled by the voltage applied to the metal over the insulator.

It is a field-effect transistor in that conductance in a semiconductor region between P-N junctions is controlled by electric force the result of voltage applied across insulator between this region and a metal electrode, *gate*. The insulator extends from P-N junction to P-N junction and slightly overlaps them. This is an insulated-gate FET, similar to that attempted at Bell Labs in the 1940s.

Khang's transistor is the result of experiments, discoveries and analysis that has gone on for many years. Of historical importance has been the work of Lilienfeld, Heils, Bardeen, Brattain, Shockley, Brown, Ross, Hoerni and Noyce. The Patent Office considers such work now to be obvious to anyone familiar with the "art". But Khang's work is sufficiently original to warrant a patent. Khang's device inspires intense activity leading to knowledge and understanding of the silicon surface oxidized to form silicon dioxide, SiO2.

The following description and operation of this insulated-gate field-effect-transistor, called a surface field-effect transistor in Andrew Grove's book, is essentially as follows:

When a voltage is applied between the gate metal electrode and an N-type *source* in P-type semiconductor, positive to the gate, electric force in the oxide insulator, forward biases the source P-N junction, for easy current flow, attracts electrons to fill holes in the P-type semiconductor near its surface and create N-type semiconductor there in an *inversion layer*. Conduction can occur at the semiconductor surface in a conductive path, *channel*, between the source and the second N-region, which is the *drain*. To the drain is applied a positive voltage to reverse bias the P-N junction between the drain and the channel. This pulls electrons along the channel and across the drain junction to the drain.

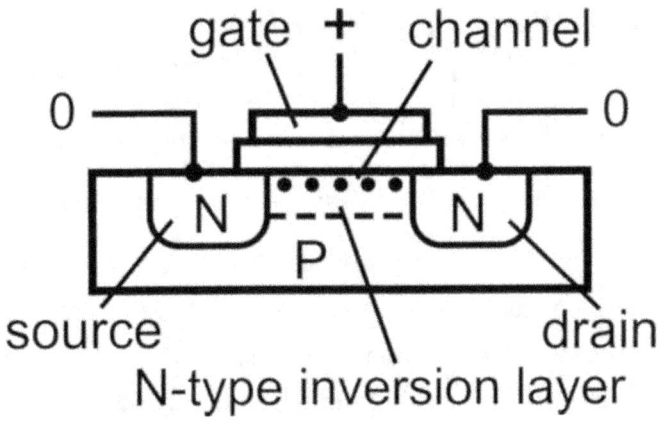

Khang's insulated-gate field-effect transistor. A positive voltage applied to the gate attracts electrons to a channel at the semiconductor surface.

The conductance of the channel between source and drain depends on the number of electrons attracted to the channel, and therefore depends on the voltage applied between the metal gate and the N-type source. When a voltage is applied and increased between the N-type drain and source, the electron flow from the source to the drain increases. The transistor is said to be in the *linear region of operation*.

But the voltage between the drain and the gate decreases and the number of electrons in the channel near the drain decreases until there is no channel at the drain, i.e., until the channel is said to be *pinched off*. As the voltage at the drain electrode is increased still further, the current changes little unless the voltage to the gate is varied. The transistor is said to be in the *saturation region of operation*.

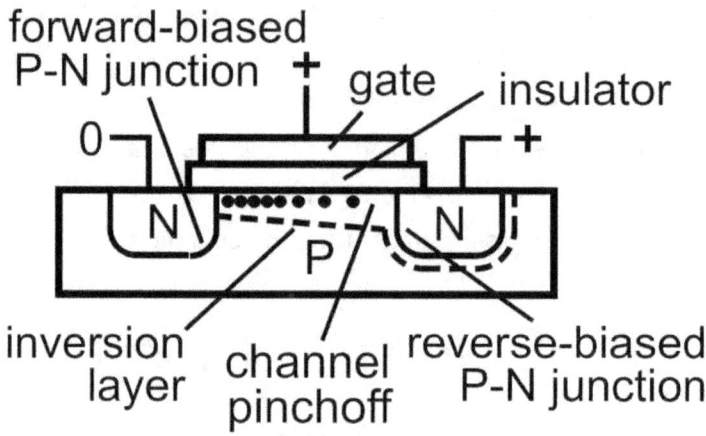

An insulated-gate field-effect transistor with channel pinched off by voltage at the drain exceeding voltage at the metal gate. Further raising drain voltage produces little increase in drain current.

A small change in gate voltage will significantly change the drain current to a typical low conductance load connected between drain and source. There can therefore be high voltage amplification. The gate is insulated from the channel by the silicon dioxide layer which is a good insulator. Leakage current across the oxide is very small. Therefore if the gate is made the input to a circuit and the drain the output, the power amplification is very high.

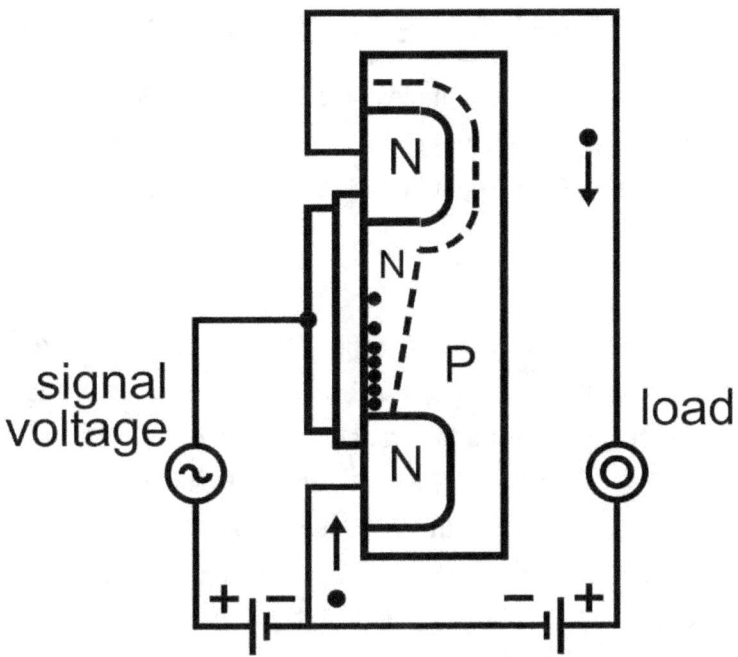

As the gate voltage in an insulated-gate FET in the saturation mode is varied the drain current changes. There is only a small gate leakage current.

The above is an N-channel transistor. The transistor can instead be processed to be a P-channel type by reversing the N- and P-regions. The channel is along a P-type inversion layer at the surface of N-type semiconductor. Electrons flow from atom to atom along the channel so that HOLES which carry positive charge flow from a P-type source along the channel to a P-type drain to which voltage has been applied, negative to the drain. Negative voltage is also applied to the metal gate.

In a paper he has presented at a conference, John Moll, who has worked at Bell Labs and is now teaching at Stanford University, in 1959 calls this transistor an MOSFET, an acronym for metal-oxide-semiconductor field-effect-transistor, and this name will stick.

MOSFET GATE TURN-ON VOLTAGE

Two major innovations are made in the mid-1960s. One is control of gate turn-on voltage in the silicon N-channel MOSFET for which, in 1963, F.P. Heiman at RCA files for a patent that will be issued in 1966.

HEIMAN

Frederic Heiman graduated from the City College of New York with an electrical engineering degree, joined RCA Laboratories under the Graduate Study Program, and received his first advanced degree in 1962. He helped develop an RCA insulated-gate FET with Steven Hofstein.

It is difficult to fabricate an N-channel MOSFET that has zero drain current at zero gate voltage. There can still be some electrons crossing the source P-N junction. By applying negative voltage to the silicon substrate, or body, in which these transistors are made, the voltage on the source P-N junction is reverse-bias at zero gate voltage, making the drain current be zero up to some gate turn-on voltage. This allows IBM to use N-channel transistors, which can be made to switch at higher speed, in their 370 computer.

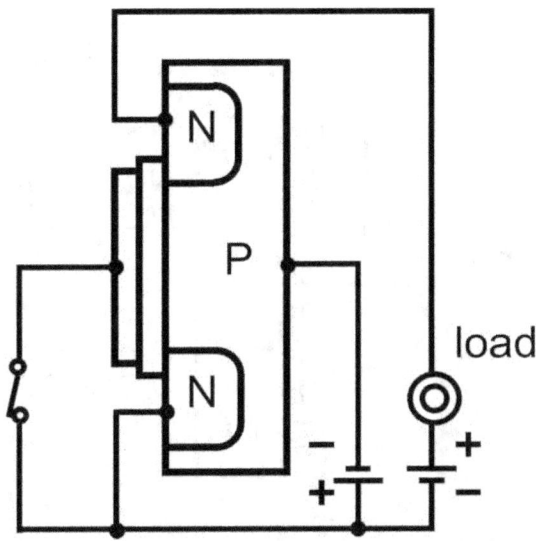

Heiman's N-channel MOSFET having reverse voltage applied across the source P-N junction to force channel current to zero at zero gate voltage (switch closed).

CMOS

In 1963 Frank Wanlass at Fairchild files for a patent on a circuit using both a P-channel and an N-channel MOSFET. The drains are connected together and the gates are connected together, so that only one of the transistors is conducting at a time during switching and neither between switching.

WANLASS

Frank Wanlass was born in 1933 in Arizona, studied electrical engineering, and received advanced degrees from the University of Utah. While a graduate student at Utah he ran an electronic circuit engineering company. In three months after joining Fairchild he conceived the CMOS circuit.

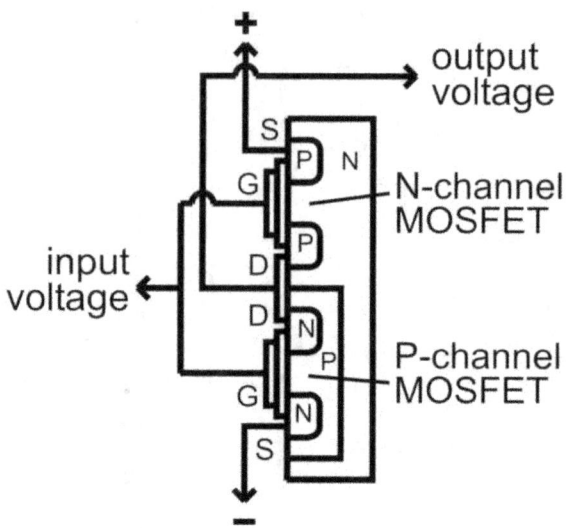

Wanlass' complimentary MOSFET circuit.

The transistor sources are connected to positive and negative supply voltages. If the voltage at the gates, which are connected together, is above a predetermined positive value, drain current can flow in only the N-channel transistor, but not in the P-channel transistor. When the gate voltage is below a predetermined negative value, drain current will flow only in the P-channel transistor. After switching, the drain current in both transistors is essentially zero, and so also is transistor power dissipation. When the N-channel transistor is conducting, it is the load for the P-channel transistor, which is not conducting. Conversely, when the P-channel transistor is conducting, it is the load for the N-channel transistor, which is not conducting. This use of both P-channel and N-channel MOS transistors is called complementary MOS, or CMOS.

It is 1964. Government sales have been taking 100% of the market. Commercial MOSFETs are announced, one for switching by Fairchild and one for amplification by RCA.

The MOSFET has appeared several times as a desirable concept. It is smaller and easier to fabricate and more devices can be formed on a given chip. Thus it has an advantage in functions per chip and becomes the choice for most integrated circuit applications.

RCA comes out with a MOSFET having a second gate between the first gate and the drain to act like the second grid in the vacuum tetrode. This transistor is used in the Scott stereo hi-fi f-m tuner.

Honeywell sets up a semiconductor design center in 1965. Westinghouse introduces a numerical controller. Fairchild cuts prices to increase demand.

Sodium ion contamination is identified as the cause of drift in MOSFET turn-on gate voltage. Silicon nitride deposited onto the silicon oxide prevents sodium ions from moving in the oxide. Also, a phosphorous glass deposited onto the oxide takes up sodium. Clean room procedures minimize contamination. Much of this work is done at IBM.

In 1966 Texas Instruments, T.I., begins work on a calculator and in 1967 Kilby, Merryman and Van Tassel file for a patent. The primary electronics of the calculator is an integrated semiconductor circuit array in one layer. There are four integrated circuits, each one of these having over 100 logic gates, all identically constructed. The NAND gates have input PNP junction transistors, two NPN transistors and one output. If five inputs AND the sixth input are at positive voltage, then the output is NOT at positive voltage, thus this logic gate is a NAND gate.

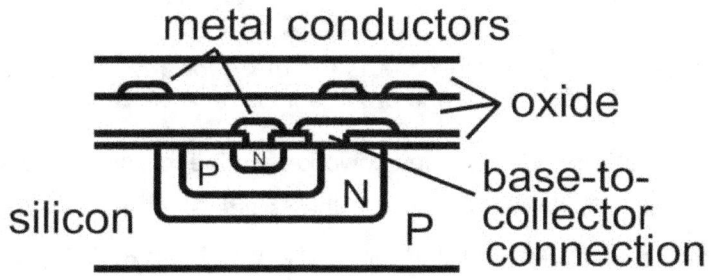

An NPN transistor in Kilby's calculator integrated circuit is functioning as a diode because of the base-to-collector connection shorting that P-N junction.. The lower oxide is grown and the upper two are deposited.

They will file again in 1972 and introduce the Pocketronic, the world's first pocket calculator, a 4-function, 2-1/2 lb, $150 calculator that will cost $10 in 1980 after sales have doubled every year.

THE SILICON GATE

As MOSFETs are being made smaller to fabricate more circuits on a chip, the relative positioning of the gate which must just overlap the source and the drain becomes critical. At Hughes Aircraft, Hans Dill in October 1966 files a patent application for a self-positioned gate. Instead of metal, silicon is deposited onto the oxide and etched leaving a silicon gate and windows through which to remove oxide. An impurity is then driven into the silicon gate and the opened windows to increase the conductance of the silicon gate and change semiconductor type for source and drain regions. This provides self-alignment of the gate between the source and drain junctions and avoids critical and difficult gate alignment problems. The deposited gate silicon is not single crystal as is the transistor body, but instead is polycrystalline, consisting of crystallites.

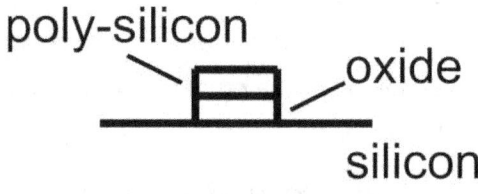

Poly-silicon gate ready to serve as the mask for driving in impurity to form and self-align source and drain regions for a MOSFET

At Bell Labs Robert Kerwin and coworkers refine this silicon gate process. They deposit silicon nitride onto the oxide for improved transistor characteristics and then a thicker oxide. Kerwin, Klein and Sarace file for a patent in March 1967.

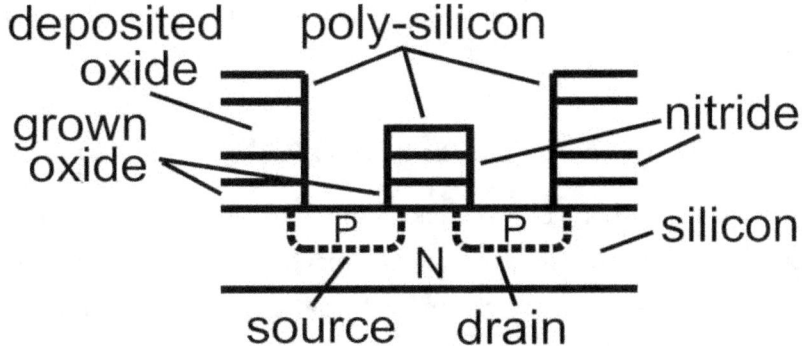

Kerwin's nearly completed silicon-gate transistor, having gate insulator of oxide plus nitride, ready for metal deposit. After impurities are driven in, the poly-silicon gate slightly overlaps the source and drain.

The self-aligning silicon gate process will become an indispensable process step. It can greatly reduce the size of MOSFETs making it possible to continue to double the number of transistors per given area on one chip of silicon in one integrated circuit, make each transistor 3-5 times faster for a given power consumption and reduce reverse-biased junction leakage current by a factor of 10 to 100. Frederico Faggin at Fairchild designs the first commercial silicon-gate integrated circuit.

FAGGIN

Federico Faggin was born in 1941 in Italy. After high school he helped design and put into use a transistor computer. In college he majored in physics and then went to work for SGS Fairchild in Italy, where he developed an MOS process and designed integrated circuits. In 1968 he moves to California and at Fairchild works with silicon gates.

It has been reported that charge can be stored for days on a transistor gate. If no direct connection is made to the conductive gate, the gate is called a floating gate. If electrons can be injected from the channel across the oxide into the gate, then negative electric charge is stored there until electrons are injected back across the oxide. Wanlass at Fairchild saw the transistor's potential as a floating-gate memory cell, but the idea was not pursued further.

FLOATING GATE

In 1967 Dawon Khang at Bell Labs files for a patent on a MOSFET having memory accomplished using a floating zirconium gate, 0.1 micron (.0001 mm) thick, deposited onto the silicon oxide and a zirconium oxide deposited over that. A positive 50-V 0.5-microsecond voltage pulse applied outside the zirconium oxide causes electrons to jump from silicon across the silicon dioxide to the floating gate, where they are trapped until a negative pulse causes them to jump back.

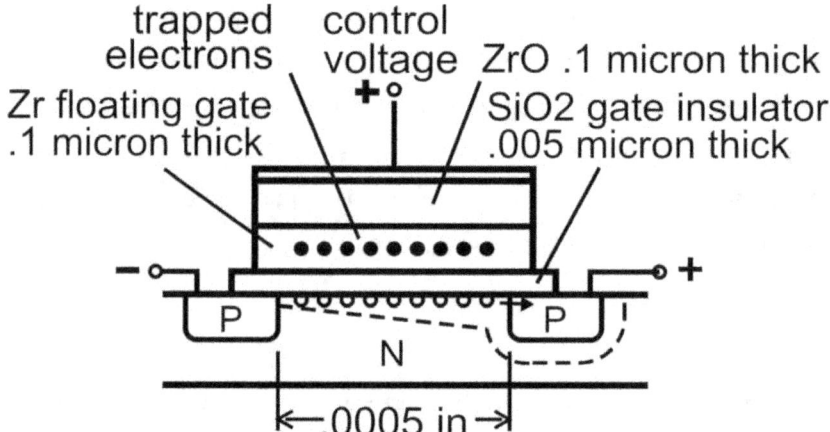

A MOSFET having a floating gate. A zirconium metal floating gate traps electrons when a positive control voltage to the gate causes electrons to jump across silicon dioxide gate insulator from N-type silicon. These electrons repel electrons leaving holes in a channel which conducts until a negative gate control voltage causes the electrons to jump back across the gate insulator.

While trapped they repel electrons from the surface of the N-type transistor body leaving holes between source and drain in a P-type channel that then conducts. There is less electric force in zirconium oxide than in silicon dioxide and the zirconium oxide is made thicker, 0.1 micron vs. 0.005

micron, both of which parameters inhibit injection of electrons through that zirconium oxide layer, causing them to be trapped.

In 1967 Robert Dennard at IBM files for a patent on a single-MOSFET memory cell for dynamic random access memory, DRAM. In a DRAM a memory cell after being read must be refreshed. The MOSFET serves as a switch to write and read the charge stored on a capacitor.

DRAM

DENNARD

Robert Dennard was born in Texas in 1932. He studied electrical engineering at Southern Methodist University. He earned advanced degrees there and at the Carnegie Institute of Technology and went to work for IBM.

The DRAM is an array of *binary* bits, i.e., a stack of words and their bits, one being addressed as columns and the other as rows. In Dennard's memory cell each cell consists of one MOSFET and one MOS capacitor. A charged capacitor acts as a binary one and a discharged capacitor as a binary zero. The transistor charges and discharges the capacitor. Word line conductors are connected to the transistor gates and turn on the transistors one whole word at a time. Bit line conductors discharge and charge the capacitors through individual transistors in the word.

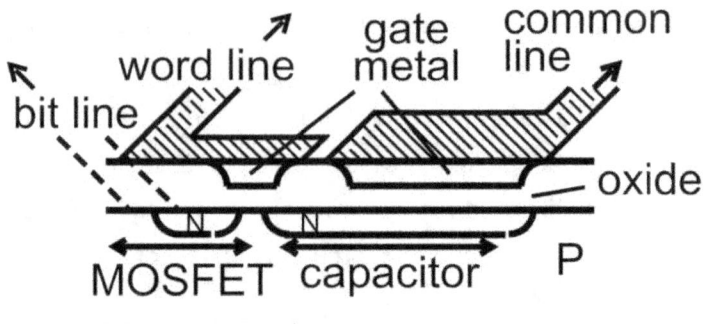

Dennard's DRAM cell. The word line conductor contacts the MOSFET gate and the bit line the MOSFET source. The MOSFET drain extends under the capacitor oxide. The common line extends over the oxide of all the capacitors. If the bit line voltage at the source is lower than the word line voltage at the gate, the capacitor can charge, otherwise it can discharge.

Profits from bipolar transistors are not sufficient for Fairchild to build an MOS IC manufacturing plant. In 1968 Noyce and Gordon Moore decide to start their own company - Intel, to manufacture silicon MOS ICs. Andrew Grove leaves Fairchild to join them.

GROVE

Andris Grof was born in Hungary in 1936. At age four he contracted scarlet fever which left him with partial hearing. When Grof was eight years old, Hungary was occupied by Nazis and nearly 500,000 Jews were deported to labor camps, including his father. He and his mother hid in shelters. When the Russian army occupied Hungary, the Russians began picking up young Hungarian men to work in camps. At age 20 Andris escaped from then Communist Hungary with another young man into Austria.

Andris made his way to New York City in 1957 where he Anglicized his name and earned a bachelor's degree in chemical engineering at City College. Andrew then went to California and earned an advanced degree at the state university at Berkeley. He then worked for Fairchild Semiconductor until he joined Noyce and Gordon Moore soon after they left Fairchild and formed Intel. Andy works as the director of engineering as they set up to manufacture RAM chips. Their first product is a 256-bit random-access memory, RAM, integrated circuit, IC, using silicon-gate MOSFETs in 1968, and in the same year they produce a 1-Kbit RAM IC.

MICROPROCESSORS

In 1968, Viatron, a startup in Massachusetts, announces what they call a *microprocessor* built from 18 custom MOS chips on three printed circuit boards for computer terminal control. At the end of the 1960s, a Japanese company Busicom contracts Intel to make custom integrated circuits for a calculator. Wang Labs introduces a desktop calculator and in 1970 starts development of a minicomputer and a word processor.

In 1969, transistor area is now 3 square microns and conductor line width less than a micron (.001 mm). Chip size is typically 0.1 by 0.2 inch. The number of MOSFETs on each chip, depending on chip size, is 100, 1000, 10,000, greater than 10,000 to several million for small-, medium-, large-, very-large- to ultra-large-scale integrated circuits.

Large-scale integrated circuits find application in wrist watches. CMOS has been used in a watch chip to divide the natural oscillating frequency of a crystal into usable frequencies. An inexpensive watch with months of battery life becomes popular. The light-emitting diode, LED, watch face is replaced by the liquid-crystal display, LCD, to increase battery life to a year or more. Large-scale integration, with about 10,000 MOSFETs on a chip having only a 0.1-inch x 0.2-inch surface area, makes possible long battery life in a watch with more functions, including a calculator.

In 1970 Intel hires Faggin away from Fairchild , Faggin having experience with silicon gates. Intel sells $1 million worth of ICs per month. It hires away from T.I. a marketing director. Due to declining demand and Japanese dumping of memory chips at low prices, Intel chooses to stop manufacture of memory chips and only manufacture computer chips.

It has become possible to make an IC that contains all of the electronic circuits necessary in a computer processor or *CPU*, central processing unit, which does the calculating, writing, and data manipulation. It is a multi-purpose programmable electronic system which processes digital input according to instructions in memory, by operating on numbers and symbols in the binary numeral system, and outputs results digitally.

The Garrett AiResearch Central Air Data Computer is flying in the Navy F-14 fighter plane. It is a set of ICs to make up a flight control system, the design of which the Navy does not allow to be published.

For the Busicom calculator Ted Hoff at Intel proposes a four-chip design, the four chips being a CPU, a ROM in which to store programs, a dynamic RAM in which to put data, and one other chip to expand input from, e.g., a keyboard, and output to, e.g., a printer. The software engineer is Stanley Mazor and Faggin is the project leader and chip designer. The silicon-gate process makes the single-chip CPU possible.

In nine months Intel has a CPU chip 1/8 inch by 1/6 inch for 12 sq mm of surface area having 2300 MOSFETs laid out on it using P-MOS performing 92,000 4-bit instructions per second for general arithmetic operations. It is the Intel 4004 IC, which is much more than a calculator, being able to carry out instructions stored on ROM chip.

Intel announces in Electronic News the 4004 as the first microprocessor. (5) It is sold to Busicom in March 1971 and is marketed to other customers in November. Busicom never produces a product and Intel buys back the rights to it.

Computer Control Corp. has arranged for Intel to fabricate an integrated circuit for their Datapoint programmable terminal. CTC gives the specifications also to Texas Instruments, who starts working on their 8-bit microprocessor chip around April 1970. This becomes the TMC 1795 IC, for which they take out patents and which they market even after it is rejected by CTC. Intel engineers finally get their 8-bit 8008 microprocessor chip working at the end of 1971 and, after CTC gives up rights to it, goes on to commercialize it in April 1972.(5)

Integration of a whole CPU on a single chip reduces processing cost. Microprocessors replace small-scale and medium-scale integrated circuits with one or a few large-scale ICs. The microprocessor ICs are produced in large numbers by automated processes resulting in low cost. There are fewer electrical connections to fail, so reliability increases. One or more microprocessors will be used in everything from small handheld devices to supercomputers.

CONTINUING SIZE REDUCTION

From 1971 through 1973, new products that appear are a portable computer from Hewlett-Packard, H.P., a calculator from Texas Instruments, T.I., cell phones from Motorola, and microwave oven controllers from T.I. and Essex International. From 1972 on, the Intel 4K-bit DRAM chip becomes widely used, using silicon gates in 1-transistor DRAM cells.

Dennard and colleagues formulate a theory for scaling down size of MOSFET chips and present it at an IEEE electron device meeting in 1972. Dimensions and voltages can be reduced without changing electric forces throughout an integrated circuit to increase the number of transistors and their speed of switching. Dennard will be awarded the National Medal of Technology and Innovation, the IEEE Edison Medal and the Charles Stark Draper Prize.

In 1975, Steve Jobs, after tinkering with electronics in a garage, dropping out of college, and designing Atari video games, with Steve Wozniak begins the Apple company in that garage. Wozniak, after building a computer as a teenager, flitting in and out of college and joining a computer club, designs the Apple I and Apple II computers. Bill Gates, after becoming interested in computers as a teenager, dropping out of college and creating and selling the computer program BASIC, with Paul Allen founds Microsoft.

In 1975-6 Intel offers a 16K-bit integrated circuit. Both the *wafer* (silicon slice) area has increased and transistor size decreased to allow for many more transistors per wafer for higher yield per wafer, lower cost per chip, and higher profit. Going to 64K-bit ICs is difficult due to expensive equipment and processing that involves new technologies.

Lower cost and more conservative Japanese designs to IBM specifications give Japanese producers the chance to eventually dominate the market. Many American manufacturers are forced to drop out.

In 1976 Wang brings out a word processor using the Intel 8080 microprocessor chip. The Apple I personal computer arrives, but Radio Shack and Commodore top it in sales of their PCs.

By 1987 DRAM cell area has been reduced to 4 square microns in an Intel 16K-bit chip. EPROM cell area is 9 square microns in a 1-Mbit (1-million bit) chip. By 1988 MOSFET delay time is reduced to nearly 4 nanoseconds (.004 microseconds) when polycrystalline silicon, refractory metal silicide and aluminum are used together in conductive material for interconnections in the 4M-bit DRAMs. Conductors are 0.5 micron wide in the 16M-bit DRAMs.

In 1981 IBM introduces its personal computer. In 1983 cellular phone services appear.

Near the end of the century in 1995, Jeff Bezos opens his garage-based business Amazon in 1995. After rigging electrical contraptions as a child, getting a college degree in computer engineering, working on Wall Street, and moving to Seattle, he opens Amazon.

In the mid-1990s the 0.35-micron generation of MOS chips is achieved, 0.35 micron being a measure of their finest features, such as gate length or contact areas. Each successive generation is assigned a number about 0.7 of that of the previous one, and since 0.7x0.7= 0.49 or about 0.5, the transistor area has been about halved each new chip generation, and so the transistor density has doubled about every two years.

In 1997 Intel packed 7.5 million transistors onto its Pentium II microprocessor chip and at the end of the century packed 42 million transistors into its Pentium IV processor chip. Intel's .13-micron generation chip which is introduced in 2001 has MOSFET gates 70 nanometers (0.07 micron or 0.00007 mm) long.

In 2001 there are nineteen companies manufacturing current generation transistors. In 2003 Intel and Advanced Micro Devices have grown, advanced and prospered. They with Hitashi, IBM, Motorola and TI offer good choices to system designers.

Transistor scaling reaches a limit in the early 2000s at about 30-nanometer (0.03 micron) feature size, where we push the limits of photolithography and await shorter wavelength extreme UV *lithography, EUV.*(7)

(As of the date of publishing this book early 2018, an Institute of Electrical and Electronic Engineers, IEEE, Spectrum article reports that EUV light having a 13.5-nm wavelength will make possible this year a chip having 7-nm features.(13) Seven nanometers is seven millionth of a millimeter. Samsung has claimed it will be ready to produce chips using EUV the second half of the year, and three competitors, Global Industries, Taiwan Semiconductor Manufacturing, and Intel are on track to do the same.(10)) And analysts think that substantially increasing the EUV power can make possible a 1-nm-generation integrated circuit. That would be a feature size only twice that of the spacing of silicon atoms in the silicon crystal.

LATER INVENTIONS

But as the dimensions of the MOSFET are shrunk to these levels, leakage currents through the transistor become a problem. At the University of California at Berkeley, Cheming Hu and his colleagues in the late 1990s develop a new integrated circuit structure that have less leakage current into the drain than the existing MOSFET structure. Their transistor is turned on its edge like a fin and is called a FinFET. This reduces leakage current by removing the silicon body from the picture and brings the gate closer to the

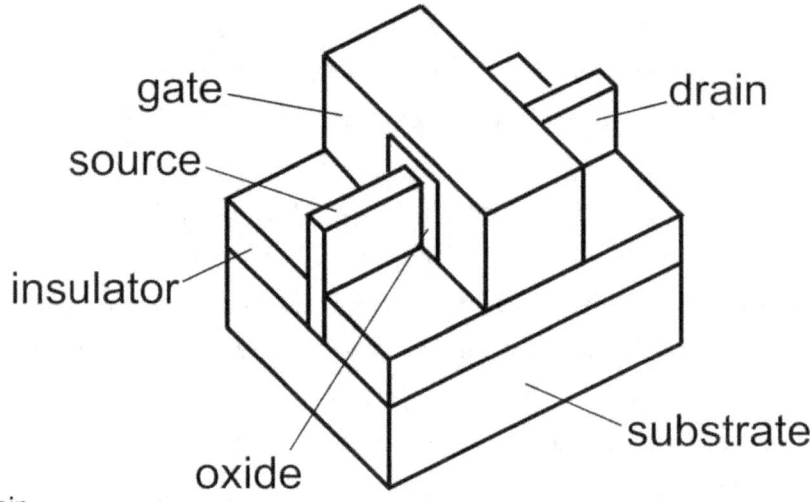

drain.

Hu's FinFET with MOSFET turned on edge for reduced leakage current. The oxide thickness must be about 10 nm.

In another three-dimensional integrated circuit, silicon wafers are stacked, interconnections being made by conducting pillars that extend down through via holes in the overlying chip.

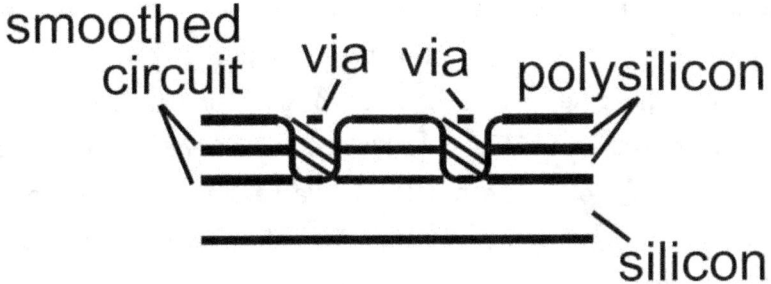

Three-dimensional chip having circuits processed on deposited polysilicon layers and connected with vias filled with conductor

Thomas Lee and Michael Farmwald, at their 1998 startup company Matrix Semiconductor, make three-dimensional circuits by depositing multiple silicon, insulator and conductor layers and, between layers, chemical-mechanical polishing each layer to provide smooth surfaces for photolithography. The silicon layers on the silicon wafer are deposited and baked at temperatures sufficiently low to not harm layers below.

The deposited silicon is polysilicon, made up of single-crystal silicon crystallites, so it is necessary to route signals around boundaries between crystallites. Later other companies will take similar approaches to three dimensional circuits, one being to bond a second wafer upside-down on the first, and stripping away all but a thin layer of silicon of the second.(8)

These three-dimensional solutions can go only so far. As the MOSFET is made smaller the oxide is made thinner to reduce the voltage change at the MOSFET gate to switch on and off the transistor. But since this thickness is approaching that of a few layers of atoms, a different approach to reduce necessary gate voltage swing is to make use of electron tunneling from source to channel. Published work on tunneling transistors appears in 2004.(9)

The transistor body can be left *intrinsic*, i.e., without the impurities that would make it N-type or P-type. This prevents electron flow from a P-type source unless the gate voltage is made positive enough for electrons to jump, tunnel, directly out of the crystal lattice in the source into the channel. A small gate voltage change can produce a large drain current change. But neither silicon nor germanium work good as the semiconductor, so other materials are being tested.

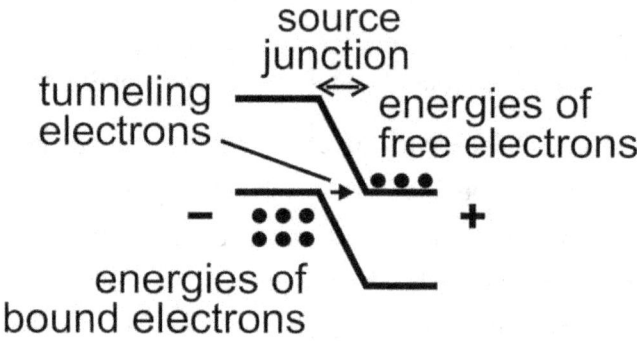

Source-channel junction in a tunneling transistor. A positive gate voltage lowers possible energy of channel electrons for slight overlap with energy of bound electrons in the semiconductor crystal.

Tunneling is an effect of quantum mechanics, which is the physics which treats very small dimensions.

CONCLUSION

Ian Ross in 1998 in the Proceedings of the IEEE points out that scientists and engineers have built on luck that nature provided.(11) It has been fortunate that silicon

- is easy to purify,
- can be produced as single crystal,
- has an energy difference between conduction electrons and bound electrons suitable for transistors,
- withstands usable electric force
- small dimensions are achievable

and that there is an oxide of silicon that

- is a good insulator,
- makes a good capacitor,
- can be used as a diffusion mask
- with a nitride of silicon provides protection from the environment.

Ross also writes that design and development of semiconductor product has been the result of largely industry efforts funded by industry itself. The military has made a major contribution by being a customer and not classifying the results. Government funding has enabled universities to contribute to knowledge and be a source of science and engineering graduates. Experimentation has been accompanied by a search for sound scientific understanding. Information has been shared throughout the industry.

In a later special issue of the IEEE Spectrum, Chris Mack will write: In the early days progress came by adding more components to a bigger chip. Over the last few decades, progress became dominated by making the transistor smaller. Now making a transistor smaller no longer makes it better and more efficient. For the last decade transistors have been made smaller to make them cheaper. But now lithography has gotten more expensive.(12)

Mack continues by writing that we have benefited from miraculous development, which shaped our modern world. We celebrate a glorious history and consider the inevitable decline. We find that the end will be rather gradual and complicated.

Gordon Moore of Intel predicted big things driven by a doubling, year after year, in the number of circuit components on a semiconductor chip. *"Moore's Law"* has described a remarkable, 50-year long winning streak. It is a testament to hard work, human ingenuity and incentives of a free market. He opened his 1965 paper saying that the future of integrated electronics is the future of electronics itself.

In the 20th century, electronics was offered to the public in radio and radio required amplification, and so:
- various coherers were patented,
- rectification of radio signals made possible voice transmission wirelessly,
- semiconductors were found to rectify,
- both electrons and holes were available,
- various forms of the transistor were invented,
- transistor action moved from bulk to surface,
- insulated-gate field-effect transistors were studied,
- and the MOSFET was developed and shrunk to ever smaller dimensions for higher speed and lower cost for new products having more and more capability.

GLOSSARY

TERM: **DEFINITION**

a-c: alternating-current, also generating or using such

acceptor: impurity atom which makes a semiconductor P-type

alternating voltage: voltage which alternates in polarity

alternating-current: current flowing in alternating direction

antenna: conductor for receiving or transmitting electromagnetic waves

Audion: Lee de Forest's vacuum triode

barretter: conductor the conductance of which varies with temperature

battery: a source of d-c electric voltage

binary: base 2 number system

bit lines: memory columns into and from which to enter and recall data

capacitor: device in and out of which electric charge slows voltage change

carborundum: a compound of carbon and silicon

cat-whisker: wire point contact to a semiconductor

choking coil inductor used to block a-c current

CMOS: MOS employing both an N- and a P-channel transistor

coherer: device that conducts when metal particles and electrodes adhere

condenser: capacitor

CPU: central processing unit in a computer or microprocessor

crystal: a sample of crystalline solid

___*single:* a sample of uninterrupted crystal structure

crystalline: having atoms fixed in a uniform repeating lattice structure

d-c: direct current, also generating or using such

decoherer: device that stops conduction in a coherer

depletion layer: semiconductor region from which free electrons or holes are removed

diode: two-electrode electron device

direct current: current which flows in one direction only

donor: impurity atom which makes a semiconductor N-type

double-diffused: semiconductor type changed twice by impurity diffusion

DRAM: RAM which is repeatedly read and rewritten

EEPROM: electrically erasable programmable ROM

electric:

___*alternator:* a generator of a-c voltage

___*breakdown:* electric discharge in solid material

___*charge:* a physical property of the electron assigned negative

___*circuit:* conducting loops interconnected

___*common:* node in a circuit from which other node voltages are specified

___*conductance:* the current that flows per unit of voltage applied

___*conductivity:* current per unit of area per unit of electric force (field)

___*conductor:* material with relatively high conductance

___*current:* electric charge flow

___*discharge:* abrupt flow of charge from a device

___*field:* electric force on electric charge per unit of charge

___*filament:* a fine wire conductor

___*force: force* exerted by electric charge

___*grid:* a folded wire electrode that controls charge flow

___*insulator:* material with very low conductance

___*spark:* electric discharge with a flash of light and a snap of sound

___*voltage:* product of electric force and distance

EEPROM: electrically erasable PROM

electrode: conductor injecting or collecting electrons at medium boundary

electrolyte: substance that makes a liquid solution a conductor

electromagnet: ferrous material wound with wire carrying electric current

electromagnetic waves: perpendicular electric and magnetic traveling forces *electron:* a point of negative charge outside the nucleus of an atom

___*spin:* a physical property of the electron that exerts magnetic force

energy

___*bandgap:* impossible energy between free and bound electrons in crystals

___*bands:* ranges of possible energy which electrons may have in crystals

___*barrier:* energy that electrons must gain to flow across some contacts

___*shells:* groups of different energy of electrons in an atom

EPROM: erasable programmable ROM

feedback circuit: circuit in which some of amplifier output is connected back to input

field-effect: effect of an electric force made to cause in a transistor

FinFET: field-effect transistor having a perpendicular fin-shaped element

floating gate: MOSFET gate to which no direct electrical connection can be made

galena: a compound of lead and sulfur

galvanometer: an instrument for measuring charge flow

hertz: unit of frequency

hole: absence of an electron in a crystal

IC generation: family of integrated circuits made using new technologies

impurity diffusion: spreading of impurity atoms into a semiconductor

induction coil: transformer

inductor: device in which magnetic force resists current change

insulated-gate: gate separated from a conducting channel by an insulator

insulator, high-k: insulator with high charge per unit electric force

integrated circuit: electronic circuit wholly contained in one semiconductor crystal

intrinsic: having the same number of electrons as the number of holes

inversion layer: semiconductor surface or contact changed in semiconductor type

Leyden jar: an early capacitor using the wall of a glass flask of water

lightning arrester: device for conducting lightning discharge to earth

lithography: the etching of windows in oxide

monolithic: formed by lithography on surface of one semiconductor crystal

Morse code: code of dots and dashes representing the alphabet

MOSFET: metal-oxide-semiconductor field-effect transistor

___*linear region of operation:*   drain current increasing with drain voltage

___*saturation region of operation:*   drain current depending only on gate voltage

NAND gate: circuit the output of which is not binary one if all inputs are one

N-type: semiconductor made by donor atoms to be a conductor of electrons

pinchoff: condition in which an IGFET transistor channel vanishes at the drain

planar process: transistor diffusion method maintaining a flat surface

P-N junction: contact between N-type and P-type semiconductor

point-contact: pointed-wire electrical connection with a semiconductor

polysilicon: made up of multiple single crystals

P-type: semiconductor made by acceptor atoms to be a conductor of holes

radar: radio detection and ranging system

RAM: random access memory

rectifier: contact across which electrons flow easily in only one direction

regenerative circuit: circuit using feedback to generate a-c voltage

relay: electromagnetic device that is a current operated amplifying switch

resonance: peaking of voltage or current at a certain frequency

ROM: read-only memory

scaling: reducing dimensions and voltages to raise density, speed and power

semiconductor: insulator with small energy gap between bands

___*chip:*   dice of semiconductor wafer

___*wafer:*   slice of semiconductor

signal: voltage or current made to vary to transmit information

solid-state: being a solid material, as opposed to liquid or gas

spark gap: gap across which an electric discharge is made to occur *syntony:* resonance

telegraph: instruments used to transmit and receive messages in Morse code

transformer: two-coil device raising current and lowering voltage or vice versa

___*primary:* input coil of a transformer

___*secondary:* output coil of a transformer

transistor: solid-state amplifier or switch

___*base:* junction transistor thin center region which controls current flow

___*channel:* MOSFET semiconductor surface along which current flows

___*collector:* junction transistor region which collects controlled current

___*drain:* MOSFET region which collects current controlled by the gate

___*emitter:* junction transistor region which emits current into the base

___*gate:* MOSFET conductor separated from the channel by oxide

___*source:* MOSFET region which emits current into the channel

trapping: interruption of electron flow

triode: three-electrode electron device

tunneling: quantum-mechanic electron transmission through a barrier

Tyranny of Numbers: size of electronic circuits reaching a lower limit

vacuum tube: device in which current flows between electrodes in a vacuum

wafer: slice to be diced

wave detector: device for detecting antenna current generated by EM waves

wave peaks: maxima and minima of oscillations of electric force

wavelength: distance between peaks of a wave at any point in time

word lines: memory rows scanned one at a time to read or write data

PATENTS

Armstrong, E. Howard. US 1113149. Filed 10/29/1913, issued 10/6/1914

Bardeen, John. US 2524033. Filed 2/26/1948, issued 10/3/1950.

Bardeen, John and Walter Brattain. US 2524035. Filed 6/17/1948, issued 10/3/1950

Bose. J. C. US 755840. Filed 9/30/1901, issued 3/20/1904.

Branly, E. US 796800. Filed 9/4/1902, issued 8/8/1905

Brattain, Walter and Robert Gibney. US 2524034. Filed 2/26/1948, issued 10/3/1950

Collins, A. F. US 644497. Filed 11/7/1999, issued 2/27/1900

De Forest, Lee. US 716000, Filed 7/5/1901, issued 12/16/1902
--------------------, US 824637. Filed 1/18/1906, issued 6/26/1906
--------------------, US 879532. Filed 1/29/1907, issued 2/18/1908.

Deitz, Burr, US 1214655. Filed 10/15/1913, issued 2/6/1917

Dennard, Robert. US 3387286. Filed 7/14/1967, issued 6/4/1968

Ducretet, Eugene. US 665957. Filed 5/22/1899, issued 1/15/1901.

Dill, Hans. US 3544399. Filed 10/26/1966, issued 12/1/1970.

Dunwoody, Henry. 837616. Filed 3/23/1906, issued 12/4/1906.

Fessenden, Reginald. US 706737. Filed 5/22/1901, issued 8/12/1902
-------------------------------, US 706738. Filed 5/29/1901, issued 8/12/1902
-------------------------------, US 706742. Filed 6/6/1902, issued 8/12/1902.
-------------------------------, US 727331. Filed 4/9/1903, issued 5/5/1903.

Fleming, John. US 803684. Filed 4/19/1905, issued 11/7/1905.

Heiman, Frederic. US 3233123. Filed 2/14/1963, issued 2/1/1966.

Hoerni, Jean. US 3025589. Filed 5/1/1959, issued 3/20/1962.

Hogg, William. US 763894. Filed 2/9/1903, issued 6/28/1904.

Holst, Giles and Wm. van Geel. US 2173904, Filed 3/4/1936, issued 9/26/1939.

Kerwin, R, E., D. L. Klein and J. C. Sarace. US 3475234. Filed 3/27/67, issued 10/28/69.

Khang, Dawon. US 3102230. Filed 5/31/1960, issued 8/27/1963.

--------------------,US 3500142. Filed 6/5/1967, issued 3/10/1970.

Kilby, J. S. US 3138743. Filed 2/6/1959, issued 6/23/1964.

-------------, US 3819921. Filed 12/21/1972, issued 6/25/1974.

King, James. US 729497. Filed 7/23/1902, issued 5/26/1903.

Kitsee, Isidor. US 657223. Filed 5/20/1899, issued 9/4/1900.

Koepfel, Adolph. US 670711. Filed 8/25/1900, issued 3/26/1901.

Marconi, Guglielmo. US 586193. Filed 12/7/1896, issued 7/13/1897.

Murgas, Joseph. US 759825. Filed 10/2/1903, issued 5/10/1904.

Khang, Dawon. US 3102230. Filed 5/31/1960, issued 8/27/1963.

Lilienfeld, Julius. US 1745175. Filed 10/8/1926, issued 1/18/1930.
---------------------, US 1900018. Filed 1/28/1928, issued 3/7/1933.
---------------------, US 1877140. Filed 12/8/1928, issued 9/13/1932

Lodge, Oliver. US 609154. Filed 2/1/1898, issued 8/16/1898.
-------------------, US 624516. Filed 1/5/1999, issued 5/9/1899.
-------------------, US 627650. Filed 1/5/1899, issued 6/27/1889.
-------------------, US 762829. Filed 6/28/1902, issued 6/14/1904

Noyce, R. N. US 2981877. Filed 7/30/1959, issued 4/25/1961

Ohl, R. S. US 2402661. Filed 3/1/1941, issued 6/25/1946.
------------, US 2402839. Filed 3/27/1941, issued 6/25/1946.

Pankove, Jacques. US 2829075. Filed 9/9/1954, issued 4/1/1958.

Pearson, Gerald and Wm. Shockley. Filed 9/24/1948, issued 4/4/1950.

Pfann, Wm., J. Scaff and A. White. US 2430028. Filed 3/16/1943, issued 11/4/1947.

Pickard, G. W. US 708070. Filed 10/29/1901, issued 9/2/1902.
------------------, US 836531, Filed 8/20/1906, issued 11/20/1906.
------------------, US 888191. Filed 3/9/1907, issued 5/20/1909.

Plecher, Andrew. US 817664. Filed 12/27/1904, issued 4/10/1906.

Scaff, J. H. and H. C. Theurer. US 2631356. Filed 12/24/1947, issued 9/18/1951.

Shockley, Wm. US 2569347. Filed 6/26/1948, issued 9/25/1951.
-------------------, US 2744970. Filed 8/24/1951, issued 5/8/1956.

Shoemaker, Harry. US 703712. Filed 6/18/1963, issued 7/1/1902.

Thompson, J. and R. US 1245135. Filed 10/18/1916, issued 10/30/1917.

Turney, Eugene. US 1144399. Filed 8/22/1914, issued 6/29/1915.

Wanlass, Frank. US 3356858. Filed 12/5/1967, issued 12/5/1967

Weber, H. C. US 1949383. Filed 2/13/1930, issued 2/27,1934.

REFERENCES

1. Dummer, G. 1970. *Materials for conductive and resistive functions.* 155.

2. Falcon, E. and B. Castaing. 2008. "Electrical conductivity in granular media and Branly's coherer. *Ecole Normale Superieure di Lyon.*

3. Alley, C. and K. Atwood. *Electronic Engineering.* 544.

4. Henisch, H. 1957. *Rectifying Semi-conductor Contacts.* 17

5. Sherriff, Ken. 2016. The surprising story of the first microprocessors. *IEEE Spectrum. September.* 49.

7. Dennard, R. 2015. Past progress and future challenges in LSI technology. *IEEE Solid-State Circuits Magazine.* Spring. 29

8. Lee, T. and M. Farmwald. 2002. A vertical leap for microcircuits. *Scientific American. January.* 53

9. Seabaugh, Alan. 2013. The tunneling transistor. *IEEE Spectrum.* October. 35

10. Guizzo, Eric. 2018. EUV lithography finally ready for fabs. IEEE Spectrum. January. 46.

11. Ross, I. 1998. The invention of the transistor. *Proceedings of the IEEE.* January. 7.

12. Mack, C. 2015. The multiple lives of Moore's Law. *IEEE Spectrum.* April. 30.

OTHER SOURCES

Wikipedia.com on early inventors, the electron, EM waves.

Lewis, T. 1934. *Empire of the Air* on Marconi.

Archer, G. 1938. *History of Radio* on radio.

Belrose, J. 1995. Fessenden and Marconi: Their Differing Technologies and Transatlantic Experiments During the First Decade of this Century on radio.

Reid, T. 1984. *The Chip* on semiconductors, the first transistors, integrated circuits.

Hemenway, C., et.al., 1967. *Physical Electronics* on metal-semiconductor contacts.

Coblenz and Owen. 1955. *Transistors: Theory and Operation* on transistor operation.

Riordan, M. and L. Haddeson. 1997. *Crystal Fire* on silicon, silicon oxide, junction transistor, Shockley.

Zygmont, J. *Microchip* on transistor and IC inventors, semiconductor products and sales.

Grove, A. 1967. *Physics and Technology of Semiconductor Devices* on insulated-gate field-effect transistor.

Sah, C. 1988. Evolution of the MOS transistor. *Proceedings of the IEEE*. p 1280 on MOSFET improvement, size scaling, applications.

###

JP

Thank you for reading my book. Please take a moment to leave me a review.

ABOUT YOUR AUTHOR

John Plumb grew up in Shelby County in Iowa and attended Iowa State College (now ISU) majoring in electrical engineering. He earned an M.S. degree from the University of Minnesota and a Ph.D. degree from New York University both in electrical engineering. After serving in the Air Force he worked with aircraft instrument test at Minneapolis Honeywell. He has held a teaching position at the University of Connecticut and worked in integrated circuit design at Transitron in Wakefield, Massachusetts. He retired in 1999 after working for Sylvania Lighting Products on solid-state discharge displays and reduced-mercury high-intensity-discharge lamps. Dr. Plumb now lives with his wife Mary in Danvers, Massachusetts and during the winter in Venice, Florida.

MORE BOOKS BY YOUR AUTHOR

The Taming of the Electron: A Story of Electric Charge and Discharge for Lighting and Electronics

Engineer: A Memoir

For information go to http://electronhistory.com

CONNECT WITH YOUR AUTHOR

Like me on Facebook: http://facebook.com/jplumbhistory

Follow me on Twitter: http://twitter.com/johnlplumb

www.ingramcontent.com/pod-product-compliance
Lightning Source LLC
Chambersburg PA
CBHW071518220526
45472CB00003B/1066